Children of the NILE

Children of the NILE

Olfet Agrama

Printed in the United States of America
ISBN 978-1-958434-58-1 (sc)

Library of Congress Control Number: 2022920240

2022.11.03

MainSpring Books
5901 W. Century Blvd
Suite 750
Los Angeles, CA, US, 90045

www.mainspringbooks.com

This novel is dedicated to the brave women who struggled through the years to throw off centuries old oppression of the Egyptian female.

They succeeded in removing the veil off women's faces in 1923 and gained for the women the right to vote and to political representation in 1953. The struggle is far from over.

I honor these women for their inspirational role and their courage. Some of these women are:

Aisha Taymur, 1840 poet, writer and social activist Nabaweya Moussa, 1907 First Egyptian girl to graduate from high school.

Hoda Shaarawi, Ceza Nabarawi, Nabaweya Moussa, 1923 first women to take off their hijab and demand equality, the right to education and the vote.

Dr. Doreya Fahmy, 1945 among the first women to get a PHD from the Sorbonne, Professor of literature at the University of Alexandria.

Dr. Doreya Shafiq, 1950 among the first women to get a PHD from the Sorbonne, feminist and journalist.

Nawal El Saadawi, 1972 founded the first legal, independent feminist organization in Egypt. Writer and activist.

TABLE OF CONTENTS

1

NADIA

I T WAS THE MAGIC moment of dawn in Heliopolis, a suburb of Cairo.
Most people were still sleeping. Nadia saw through her window
the clear blue sky shining with the first rays of the sun. The haunting
call for the early morning prayer sent shivers down her spine. She opened
her bedroom door slowly and tiptoed barefoot on the cold marble floor into
the sitting room.

Even though she had timed it perfectly, she knew that she was taking
a very big risk. The house was quiet except for the faraway sound of the
muezzin calling people to prayer. She saw with relief that the door to her
parents' bedroom was closed. Nadia's slim young body was shaking in her
flannel nightgown, thinking of her father with dread. In the semi-darkness,
she threaded her way between the pieces of furniture and almost cried out
when she hit her shin on the corner of the piano.

Finally, she reached the phone and her fingers automatically dialed the
number that had become engraved in her memory. Ever since she had met
Sammy at the neighborhood bookstore she couldn't get him out of her mind.
His Clark Gable smile haunted her every moment.

"Hello Sammy. Are you awake?" She said trembling from the excitement
and the cold. Her father didn't believe in heating the house.

He claimed that it was unhealthy even though at night, in the desert of
Heliopolis, the temperature often dropped thirty degrees.

"Yes, my angel, I'm studying and waiting for your call," Sammy said in his low seductive voice.

"I'm so afraid that I might get caught," she said, still shaking. Was this love or was she coming down with a cold; her nose was beginning to run.

"Don't worry, they must be fast asleep. Give me a kiss."

She giggled and gave the phone a noisy kiss. She felt her face flushing and a strange sensation between her legs.

"Hey, have you made up your mind?" said Sammy. "When am I going to see you? I can't live like this anymore. Please sweetheart, I want to see you soon or I'll go crazy."

"I'm trying to find some way of skipping school but it seems a very difficult proposition," she whispered.

"Where there's a will there's a way."

She was silent, feeling suddenly tired of hearing the same dialogue night after night.

"Don't you love me?" Sammy said.

"You know I do. Don't be so impatient."

"Impatient? We've been talking almost every night for six months now and you call that impatient."

"What can I do? It's enough I'm talking to you on the phone. My father would kill me if he found out. Did you know that one of my cousins was caught talking to a neighbor and was sent to a Catholic boarding school?"

"Nadia, I have at least another year till I graduate from medical school. Let's take a chance. I want to hold you and kiss you. I can't wait another year. This is ridiculous."

"Shush. I hear a door opening. Bye." Nadia put down the phone as softly as she could and ran to the refrigerator in the nearby pantry to get a glass of water.

Her father shuffled into the pantry. He was half asleep. "Who were you talking to?" he growled.

"Nobody Papa." Nadia said softly, hiding her face in the refrigerator.

"Look at me." He raised his voice suddenly and whirled her around.

"I was talking to Melanie," she said. She could barely steady her shaky voice.

"Liar." He slapped her so hard that she thought he had broken her jaw. Tears came running down her cheeks. "Go back to bed, it's five o'clock in the morning and don't even think of going to school tomorrow." He pushed her roughly into her room and locked the door behind her. "I'm going to have a talk with your headmistress. You're going to behave or else no more school. I'll not have you disgrace me.

You hear me?"

"What's happening?" her mother asked, looking terrified as she came out of her bedroom.

"Go back to bed you stupid woman. You better watch this girl until she gets married. I don't want any nonsense in my house and she is not to talk to this Melanie friend of hers at any time. She's your responsibility, remember that."

Nadia heard her mother mumbling and slinking back to her room.

She sat down on her bed and started to cry in pain and anger. Her mother was never going to support her. That was certain. Well, she stopped crying, I'm going to elope and marry Sammy whether they like it or not. I will show them that I have a will of my own.

Nadia spent a whole week locked up in her room. She cried a lot and refused to eat but her father insisted on keeping her at home. Finally, her mother was ordered to go to the school and talk to the headmistress and insist that her daughter be watched closely and not allowed to leave the school until the family driver came to fetch her. Her father finally relented after the headmistress assured him that Nadia would be safe in the school and that his request would be granted. He reluctantly allowed her to go back to school.

The heat in the school courtyard was oppressive; a thick veil of dust covered the tall eucalyptus trees. All the students in the English School for Girls were dressed in their drab uniforms of beige and brown. They were either playing basketball or talking in clusters on the yellowing grass. The school field, like the rest of Cairo, was eternally covered with a film of brown dust.

During the lunch break, Nadia felt lonely and a little bewildered.

In spite of her father's wrath, she decided to skip school anyway and meet Sammy. She had to work out a plan with her best friend Melanie.

She walked across the rectangular grassy field in search of a shady spot where she could eat her lunch. A very high stone fence surrounded the schoolyard. The only thing visible above the wall was the top of the metro train as it whizzed by. The school was like a prison. Nadia was not sure of herself. She was confused about everything. It also wasn't clear in her mind if she was cross-eyed, bowlegged, dumpy, and ugly as her mother repeatedly told her, or if she was a beauty queen who looked like Elizabeth Taylor, the American movie star, as her school friends always told her. Her Dada adored her and told her every day that she was as beautiful as the *Hooreya's* in heaven, but she didn't believe her nurse.

She wasn't even sure what a *Hooreya* looked like. In the Koran they were supposed to be beautiful maidens who comforted good men in heaven. There was no mention of male angels loving saintly females.

There wasn't even any mention of women going to heaven! Where did the women go? Nadia wasn't sure about that either. She would have to find out. The Koran was a real mystery to her. She had to memorize some verses but she did not understand the meaning of any of them.

She finally reached the shade of a huge eucalyptus tree and sat down to eat her lunch.

Nadia saw Melanie walking rapidly towards her across the wide school field. Unlike most Egyptian girls who undulated and swayed, Melanie walked fast like a boy. She was a pale willow, skinny, and fragile.

Her wrists and clavicle bones stuck out as if she were only temporarily connected together. Her brown uniform looked faded from the sun and frequent washings. Her shoes were scruffy. She had a bobby pin holding her short, straight, blond hair away from her face. It always impressed Nadia that though Melanie was half British and looked foreign, she could swear in Arabic like an Egyptian truck driver. Although she was already nineteen, she looked like a sixteen year old tomboy.

Melanie dropped onto the grass next to Nadia and slowly unwrapped the newspaper that was holding her lunch. "What's up, Nadia?"

"I've made up my mind. I'm going to meet Sammy in secret and you are going to help me." Nadia confided to Melanie; she was blushing and her heart was beating with excitement.

Melanie looked disturbed, "Are you sure you want to do this? Damn it, your father has the school watching you."

"I don't care. I'm not a child. I'm almost eighteen. Besides, Sammy loves me and we're going to get married as soon as he graduates."

Melanie smiled, "Well, he sure has you mesmerized, silly romantic.

Has he proposed? Did he kneel down on his knees and ask you to marry him?"

"I know that he wants to marry me. He promised that he'd talk to my father as soon as he graduates. I know he will."

"Sammy is bad news."

"Why is he bad news?" asked Nadia.

"He's what we call a ladies' man; he's always surrounded by girls." Melanie said.

"Well, he is smart and handsome. What do you expect?"

"I just don't think he'll make a very good husband."

"I love him and he loves me."

"Loves you? He just wants to get into your pants."

"Melanie, don't be vulgar."

"Oh Nadia you are so innocent. Stay away from Sammy."

"Why?"

"Bloody hell, you want to know why?" Melanie could never bring herself to tell Nadia that she once had sex with Sammy after a drunken party. It didn't mean anything to either of them, still she couldn't tell Nadia. She had to change the subject.

Nadia couldn't help feeling that Melanie knew a dark secret about Sammy that she was trying to hide from her. She would soon find out anyway. There were no secrets in such a small community as Heliopolis.

"Tell me, please," she begged.

"Nothing, really serious." Melanie was silent for a moment. After thinking for what seemed an eternity, she took a deep breath, she said, "The party at the Heliopolis Club was great."

Why was Melanie trying to change the subject thought Nadia? She seemed very tense.

"The dancing lasted till midnight," Melanie continued, looking down at her sandwich. Nadia kept quiet, still hoping that Melanie was going to tell her something about Sammy.

"Sammy was there," Melanie suddenly blurted out.

Nadia felt her face turning red and a shot of jealousy went through her. She became very angry. "You were going to keep this a secret?"

"Nadia, I didn't want to hurt you."

"Who was he with?" she almost yelled. She became sick thinking of Sammy dancing at the club while she was locked up at home in her bedroom.

"Nobody special," Melanie lied, for Sammy was dancing all night with Zizi, the notorious merry widow.

"He didn't tell me, the skunk. He pretended to be suffering the agonies of love and all the while he was at a party, dancing." A look of pain and disappointment was etched on Nadia's face.

"Nadia, you can't expect Sammy to stay at home waiting for you. He's a man and men have freedoms we're not allowed."

"But you went to the dance."

"Yes, but I'm half British and all foreigners are considered whores."

"Melanie, you say the most outrageous things," said Nadia laughing bitterly. She was still angry with Melanie. "I think I'm going to lose him if I don't agree to see him soon.

"Maybe it would be better if you lose him," said Melanie.

"How can you say such a thing? I really love him."

"Love's nothing but trouble. A pain in the ass," said Melanie.

"Who was he with? Please tell me."

"He didn't have a special date. I told you that he's very popular with all the girls," said Melanie.

"You're so lucky to be able to go dancing. Whom did you dance with?" Nadia asked.

"Not with your Sammy. Just with the usual blokes. Nobody very exciting," said Melanie.

"Whom did you go with?" Nadia was curious and very envious.

"A friend of my mother's and her two daughters," said Melanie looking very uncomfortable.

Nadia's face was pale and a look of pain was clear in her eyes.

"Okay, stop giving me those tragic looks, I give up," said Melanie.

"I'll help you see him since this meeting is inevitable and you are doomed. Would you like to seal the bargain by exchanging lunches?"

Melanie said with her extra large smile.

"Oh, thank you, Melanie. You're a real friend," said Nadia happy.

She quickly exchanged her cellophane wrapped sandwiches with the newspaper package Melanie handed her. She found herself slowly crunching a sardine and tomato pita sandwich; she nearly gagged on it.

Disgusted, she looked at Melanie. "Interesting sandwich." The greasy texture of the fish was hard to swallow, and all those bones were like pins in her mouth.

Melanie laughed, guessing at Nadia's discomfort, "Eat the bloody sandwich, darn it, it's supposed to be good for you. My father is a nutritionist, the bastard should know." Melanie continued, as she enjoyed Nadia's roast beef and mayonnaise sandwich on white bread.

"Sardines with their bones give you calcium, plus unpeeled vegetables. It's very healthy, you know?"

No, Nadia didn't know. She had never eaten whole sardines nor even heard of a nutritionist before. Anyway she didn't care; her heart was breaking. "Is your father very successful?" Nadia tried to make conversation in order to hide her malaise.

"Successful?" laughed Melanie. "No, I wouldn't say that. Golly, he says he's ahead of his time but most people think he's crazy."

"Is he?" asked Nadia, admiring her friend's disrespect for her father.

"Crazy?" Melanie rolled her eyes, "Darn it, yes, he definitely is."

While they were laughing, a small black hawk came sweeping down from the sky and grabbed Nadia's sandwich from her hand.

"Oh no, not again," Nadia said, afraid to show her relief at the magical disappearance of the disgusting sandwich. The sky was full of the familiar black hawks and crows circling over the playground. They were the daily hungry visitors.

"It's lunch time for the birds," sang Melanie. The girls giggled and shared what was left of their lunches. This time they screened the food with their arms from the birds circling overhead.

"Melanie, how are you going to arrange the meeting?" Nadia said.

"God damn, I don't want to be responsible. I told you he was bad news."

"Melanie, you promised. You even ate my sandwich. Look, my parents are now very suspicious. They are watching my every move. My father forbade me from talking to you on the phone. Could you please call Sammy and tell him that I can't talk on the phone anymore? Maybe he has some ideas. I need to look at him, to touch him, and make sure that he loves me."

"Okay, I'll call him. I'm sure he'll think of something. Come on, we're late; it's time to go back to class. I'll race you," said Melanie as she took off. "Last one in class is an old maid," Melanie called back from a distance.

Nadia knew even before she started running that she was destined to become an old maid. Melanie was so fast; she couldn't keep up with her. She was quickly out of breath and it took her a long time to find her way to the class through what seemed to be an endless labyrinth.

As Nadia walked in slow motion toward the back of the class to where Melanie was already sitting, she suddenly saw clearly how little her friend fit in. In a sea of curly brown hair, Melanie's blonde head stuck out. Is that why Melanie had no other friends? Is it because she was so different? At that moment, Melanie turned to Nadia and smiled.

SAMMY

S AMMY SAT AT HIS desk thinking about Nadia. Facing him on top of the pine bookcase that he had made himself were several human bones, a skull, a small radio on which he listened to the BBC, a collection of mystery novels and his numerous medical books. He turned on the green glass lamp on his desk and stared at the bleached skull that stared back at him.

Nadia was too young and innocent to realize how risky it was to be seen alone with him in public but the danger in the situation only increased his excitement. Did he really love her or was he being impulsive? He had to see her in private. She would be a perfect wife for him. She was beautiful, she was rich and she was a virgin. He was sure that she loved him and would be a docile obedient spouse. Best of all, with her father's money he would be financially secure and socially established.

He smiled as he stared at the American movie posters facing him.

He had stolen them off the walls of Cinema Radio in Talaat Harb Street. One was of Orson Wells in *The Third Man* and the other was of Ingrid Bergman and Humphrey Bogart in *Casablanca*. Movies were his passion and these two films were his favorites.

The high pitched call for noon prayer was blaring from a nearby minaret as Sammy later walked away from the dilapidated apartment building where he lived and crossed the wide Sesostris street. The summer heat was intense in spite of the blossoming jacaranda trees that shaded both sides of the street. The purple flowers were exuberantly festive but there was positively nothing to celebrate in the aftermath of the military defeat by Israel. He could hardly breathe thinking about the political situation, and now the omnipresent dust was starting to trigger another asthma attack in his chest.

The large grass divider in the middle of the road seemed like an inviting cool oasis. He found himself seeking refuge there from the heat and the dust. The pale green strip of grass was a haven for itinerant vendors of fruit, juices and kebob sandwiches. Sammy had eaten the sandwiches for many years, until one day the police discovered that they were made of dog meat, and the vendor ended up in jail. The grass was also a playground for the children of the area, and a grazing ground for the skinny goats and sheep herded into Heliopolis by sinewy Bedouins from the Sinai desert. It was also a cool prayer zone for believers caught on the street during the hour of prayer.

He spotted some young men praying under the shade of the trees.

Sammy marveled at the calm of their faces. Like the rest of the city, there was no sign that the country had just lost the war with the newly created Israel. He thought that he could smell the blood of the Palestinians massacred recently at the village of Deir Yassine, yet life went on as usual and no one could have guessed that Egypt was simmering with frustration and smarting from defeat. He predicted that the country's anger would boil over at any moment. After thousands of years of enslavement, was there no limit to the endurance of the Egyptians?

Sammy walked fast, for in spite of his asthma, he had been a long distance runner in high school. He was thinking all the time of his new conquest, the beautiful Nadia. She was so beautiful it hurt his eyes to look at her, and he was determined to have her.

He passed Everyman's Bookstore, the only English bookstore in the quarter, where he had been buying his books and magazines and pocketing erasers and pencils since he was a boy. The store looked shabby and the dusty bookshelves were half-empty, but now it seemed a magical place because it was where he had met Nadia for the first time. For almost a year now they had been only talking on the phone or stealing a rare kiss in secluded corners at the Heliopolis Sporting Club.

He now had a plan in motion. He was on his way to borrow his father's new Buick that was parked in a public garage two streets down from where they lived. The car was essential to his plan.

Sammy was glad that his father drove his car only on Fridays and holidays. An army car and driver picked the General up in the morning at nine and brought him home at four o'clock sharp in the afternoon every day. He was lucky that General El Hakim was punctual and constant in his routine. Sammy walked hurriedly into the huge garage and waved to the guard on duty.

"*Sabah el kheir ya doctoor*," the man said in a gruff deep voice.

"*Sabah el kheir ya Ibrahim*, good morning," answered Sammy, returning the greeting. "I'm not a doctor yet, old man." He laughed as he gave him one of the ten-pound notes he had won the night before playing poker with his friends from medical school. Sammy was a good player. He almost always won even when the cards were not marked.

"Thank you, thank you. May God keep you Doctor and make you prosper," said the skinny old man, bending over with his right hand on his heart to show his gratitude. Sammy searched the man's face for a second.

"You know, Amm Ibrahim, I need a favor from you. I need the car for a few hours." A look of panic appeared on Ibrahim's face.

"I'll get into trouble." said Ibrahim, "The General will kill me if he finds out."

"Listen uncle Ibrahim, I swear that my father will never know." Said Sammy, laughing, "unless you tell him, of course." He smiled his most charming smile and put a reassuring hand on the man's shoulder. "It'll be our secret and I'll make it worth your while."

"You're tempting me master Sammy, may God forgive you."

Sammy patted Ibrahim on the back. "Don't worry, I promise you he'll never find out."

Ibrahim shook his head and limped slowly away to get the keys.

He was trembling and almost in tears but one pound was enough to feed his family for a week. Maybe now he could see a doctor about the pain in his legs.

Sammy looked hungrily at the glistening cars in the garage. The shiny curves of a Rolls Royce parked close by were as exciting as the curves of a naked woman. He went over to his father's black Buick and lovingly caressed the shiny elliptical chrome vents. The black Buick was the only luxury Dr. Hakim had reluctantly permitted himself after much nagging from his wife, Laila.

Ibrahim returned limping all the way from the dark belly of the garage. "*El Shaitan Shater*," he said. "The Devil is clever," he repeated under his breath, as he reluctantly handed the keys to Sammy.

"*Shukran Amm Ibrahim*," Sammy laughed happily. He was careful not to get his well-pressed white pants dirty from the dust and grease all around him. He got into the car and drove out of the garage, yelling, "I'll be back in a couple of hours, I promise."

Sammy reached Nadia's school in the Abbasseya district in thirty minutes. The girls were in the classrooms and the imposing Nubian guard was sitting as usual at the school gate reading the paper. "Good morning," he said to the black giant.

"*Sabah El Kheir*," answered the guard, smiling at him. His brilliant white teeth sparkled in the sun.

"I need a favor, uncle."

"My name is Osman," he said.

"Uncle Osman, I need a favor." Sammy put a one-pound note in the man's hand.

"Are you bribing me?" The man said with a scowl that would have scared a less determined man.

"May God forgive you. I just need to give a message to my cousin, Nadia Soliman. We are engaged, you see. Just a note from my mother to her."

"Nowadays young people are full of mischief," said Osman, shaking his turbaned head as he took the note.

Sammy came to the school almost every day after that and talked with the guard. He gave him cigarettes and sometimes money and once in a while gave him a note for Nadia. He knew he would get his way.

He would wear her down.

MELANIE

IN THE SUMMER MONTHS, Melanie preferred to walk the two miles from where she lived in Abbasseya to her school. The public buses were overcrowded and foul smelling, and it was always more interesting to watch the people going about their business on the street.

Still, the unbearable heat radiating from the melting asphalt streets, the diesel fumes, and the omnipresent dust gave her a headache. As she passed the butcher shop, she quickened her step. The carcasses of slaughtered animals, hanging on hooks and covered with flies, always scared her.

Semsem, the butcher's young assistant, ran out of the shop as he saw her pass. His apron was bloody, his eyes were crossed, his face was pale and spit was coming out of his gaping mouth. He dragged her into the shop and pushed her into the meat vault and bolted the door. The gory sight of raw flesh and the pungent smell of blood revolted her. She was very scared. She screamed and pounded on the door with her fists. Her hands and her face were dripping with blood.

She woke up with the bed sheets twisted around her damp body.

The only sound of pounding was that of her heart. Was she going mad? It was time to get up and go to school.

As she walked on the wide sidewalk shaded now and then by giant fichus trees, she was savoring the ripe yellow guava fruit that Ali Abbas, the fruit vendor, had given her. The perfume of the fruit was intense and the flesh of the guava was like the body of a baby, firm but incredibly soft. It tasted sweet but had a tart edge to it. The sun had ripened it to perfection. Melanie was enjoying eating the fruit slowly but she was also careful to look down and watch her step. There were always traps on her path, holes in the pavement that could easily break her ankle. Large stones left over from construction sites, discarded pieces of furniture thrown from windows, and the odd uncollected garbage spilling from the doorways.

She didn't trust Sammy but she also hated the fact that Nadia was kept prisoner like a bird in a golden cage. Was she doing the right thing in helping her skip school to meet him? She was confused and frightened. Her own experience with love had ended in complete disaster. She had no illusions about Kamal, the young man she had met during the summer vacation in the seashore town of Balteem. She was in love with him but she didn't dream of marriage, for she had no dowry or good family connections; nor did she dream of eternal love. She was a realist and surely enough, Kamal's love cooled off quickly and he was now engaged to a rich relative of his.

The Ramses juice shop was open and filled with customers. There were large glass jars of liquids of every color displayed on the counter.

The vendor was putting long sticks of sugar cane through the rollers of a hand-operated press. He gave a customer a large tumbler of frothy sugar cane juice. The flies that swarmed around the spilled juice didn't seem to bother anyone at all. Melanie would have loved a glass of sugar cane juice, or better still, some mango juice, but she knew that it was a sure way to get hepatitis, or typhoid, or even cholera. Last year's cholera epidemic had been very scary and should have taught everyone to be extra careful, but most people seemed already to have forgotten all about it.

One morning, she had asked her father for some money for new shoes. Her old worn-out shoes were hurting her feet, and her efforts at polishing them were futile. The leather was peeling and the sole was so thin she could feel every pebble in the road as she walked. He had scowled at her and said he could not afford new shoes at the moment.

Ever since he remarried, he was treating her like an unwanted burden and was always complaining about how much it cost to take care of her.

"Then, send me to England" she had said.

"Your mother doesn't want you."

"That's not true. Why are you so mean?" Melanie tried hard not to cry.

"Just finish high school and get yourself a job."

She was going to England whether he liked it or not. The ticket didn't cost very much. She would have to work and save the money, but sooner or later, she would go. Reema, her stepmother, wasn't a bad person. She was only twenty-one, and was overwhelmed with her wifely duties and the new baby. She treated Melanie like a servant because she thought that was the normal way to treat a stepdaughter. Melanie barely had time to do her homework after she finished the washing, ironing, and other chores.

Nadia was right to tell her that she was a regular Cinderella, except that in her case, she knew it would be difficult to find Prince Charming.

It was easy to find a lover - the boys were crazy about her, and she did not believe in bartering her virginity for an offer of marriage. She had a reputation for being an easy lay. No one would want to marry her now, for in Egypt a girl with a past and no dowry was not very desirable as a wife. It wasn't important; she didn't believe in marriage anyway.

Melanie wished that her Mom had taken her to live in England, but she couldn't leave the country. She was her father's property, the law of the land said that she would have to wait until she was twenty-one to be free of his control. She had to accept her fate with grace. In the meantime, after she finished high school, she would have to learn how to type and become a secretary or something. College was out of the question. Being a secretary wasn't so bad. It was better than ending up in an arranged marriage like the majority of her classmates. Working was a much better fate than being trapped in a loveless marriage.

"Good morning to the roses and the jasmine! A morning of cream and honey!" Semsem called enthusiastically from the front of the butcher shop. She panicked and quickened her steps to get away from him. She must locate a book in the library on how to interpret dreams in order to find out the meaning of her nightmare. Even though Egypt was a different world with special nightmares all of its own, she might try to understand why her dreams were so violent.

Kamal suddenly drove up and stopped along the side of the road.

"Come on I'll take you to school," he said.

She jumped in beside him even though she knew that she was being foolish.

"How've you been?" she asked.

"I missed you sweetheart."

"I like your engagement ring," she said. "Oh Kamal how can you pretend that nothing has changed?"

"I'm serious. I'll break the engagement."

"Don't tell me stories. I don't believe you."

"Can I see you tonight?"

"No, I will see you when you are a free man. I don't want to be your secret mistress."

"You are the mistress of my heart."

"Let me down at the corner. There's no need for me to be seen getting out of your flashy car."

"Shall I come to get you this evening?"

"No, Kamal. I have homework to do. Bye."

She should stop thinking about him, and yet, why not? She knew that he loved her in his own way and one never knows what could happen in life. Love is sweet, even if it is a hopeless love.

When Melanie arrived at the school, Sammy was waiting for her at the gate. She was not thrilled. He greeted her with a smile. "Could I have a word?" he said.

"Hi Sammy, what brings you here?"

"Your friend. Can you help me see her?"

"How can I do that?" Melanie was nervous. Many of the girls were giving her curious looks.

"Melanie, please just tell her to come out during recess. I'll be waiting for her tomorrow at noon."

"I don't trust you. You are very cheeky to ask me, of all people."

"It was fun being with you but this is different. I am in love with her."

"I really don't trust you. Besides it is too risky, she might get expelled."

"Don't worry Melanie. Please I will just take her to lunch, just an innocent lunch. I promise." He gave her a note for Nadia.

"You must know that I don't like what is happening. Please, I beg you, leave the girl alone. She's just a child."

"Melanie dear, my intentions are honorable. I assure you that I want to marry her. I just want to be able to go out with her once in a while.

It's not normal to have this long distance relationship. We have at least another year to wait. Please have her come out with me during recess and I will return her to school in a couple of hours. No one will know.

I promise that I will take care of her."

"I'm not sure this is a good idea. What about the guard?"

"He's okay. He's a friend of mine." Sammy said smiling.

"Of course. I should have known. I'll tell her of your plan if you promise me to respect her and take care of her."

"Cross my heart and hope to die," he said smiling.

Nadia was already in the schoolyard waiting for her. Melanie never failed to be surprised at her friend's beauty and grace. Her shiny brown hair, her dimpled smile, her elegant movements were lovely; everything about Nadia was attractive. She was motioning to Melanie to come over. She was looking fresh and rested; her clothes were crisp and clean.

Melanie felt a stab of jealousy. The dwarf was standing behind Nadia motioning to her to come closer. Melanie suddenly stood still in her tracks and couldn't move. She beckoned to her friend to approach.

"What's wrong with you?" said Nadia, as she hurried over to where her friend was standing, rooted to the ground.

"Nothing," she laughed. "Just a test to see which one of us has the stronger character," said Melanie. The dwarf disappeared. "I just read it in a book, by *Gaylord Hawser*. You came over to me; that means I have control over you." She smiled as she saw consternation on Nadia's face.

"You're really crazy. I thought something was wrong with you," she said. Melanie winced. Was she really crazy? She wondered.

"Don't say that Nadia. It makes me nervous. I sometimes feel that I am crazy."

"You're so weird sometimes," said Nadia, shaking her head. They both walked towards the classroom.

"Of course I'm weird, more weird than you think."

"What do you mean?"

"I hear voices and I see strange people."

"What do you mean?" Nadia asked, looking worried.

"Please, you must promise not to tell anyone. It's a bloody secret."

"Of course Melanie, your secret is safe with me."

"Well Nadia, there's a beastly dwarf who follows me everywhere and talks to me." Melanie said in a low voice.

"A dwarf, like the seven dwarfs of Snow White?"

"Not exactly. He is very ugly and vicious. He drools and slobbers and his face is all wrinkled and pockmarked. His nose is too big and his lips are blue and thick. His teeth are crooked and yellow and he has a bad smell. He also has a hunchback and bowlegs. He has a sly smile and a chilling laugh. He watches me all the time."

"He sounds really scary, but maybe it's all in your mind, Melanie."

"No, look behind you. He's hiding behind that tree."

Nadia looked behind her but could see no dwarf lurking in the schoolyard. "Melanie, are you joking? There's no one."

"He's here, I assure you. He follows me to school every day."

Nadia felt the hairs on her arms rising with fear. Melanie's eyes were full of terror. She was being strange. Nadia tried to calm her down.

"You know Melanie I also always dream of three girls in white.

They crouch by my bed and sometimes pull the covers off me when I'm asleep. They're always there and they scare me. My mother says it's my imagination. They never show up during the day so it must be a nightmare."

"I see him now. I see him during the day lurking in corners," said Melanie.

"You just have a wild imagination."

"The doctor says I might be schizophrenic. I hear voices all the time. He is giving me a medication."

"Melanie, you are so smart. The doctor must be wrong."

"Smart has nothing to do with it. I don't think I'm crazy. Just a wild imagination." They both laughed.

"By the way, Sammy sent you this." She gave Nadia the folded note.

"Another note. He never gives up."

While Nadia was reading the message Melanie looked wistfully at her friend's new shoes with envy. It made her sad that life was unfair but she smiled at Nadia and promised not to let herself become a bitter person.

"Okay, it's settled, tomorrow I'll walk out during the lunch recess.

He'll be waiting for me. Oh, Melanie. Thank you. I'm so excited."

IS CAIRO BURNING?

FRIDAY, JANUARY 24, 1952, around noon and in spite of her ambivalence about meeting Sammy in secret, Nadia found herself waiting for him in front of her school.

"I hope we haven't made a terrible mistake," was the last thing Melanie had said.

"Don't worry, I'll be fine," but now Nadia wasn't so sure. It wasn't too late; she could hide behind a tree and slink back into the school.

It was a glorious sunny winter day. Nadia glimpsed the shiny black Buick arriving around the bend. Her stomach muscles tightened and she started to tremble. Sammy stopped the car and jumped out as soon as he saw her and rushed towards her. She shivered slightly as he stood close looking at her with an eager smile. The sight of him caused a now familiar stir in her blood.

During the past year, she had met him at the club for a few stolen minutes. They had only time to hold hands briefly, and the last time they had met he had kissed her. He helped her get in the car.

"You've lost weight," she said.

"I have been studying all summer. I always have to take the make-up exams. Not too smart, I guess." he laughed.

"Too much partying, I think." Her legs were unsteady. She was sure that this must be love.

"Finally, I have you to myself," he said smiling.

"I have to be back in two hours." She tried to look calm.

"It doesn't give us much time. What do you want to do?" he said as he drove off. Before she could come up with any suggestions, he said, "I know, let's go to the zoo."

Nadia was disappointed. She suspected that he was taking her to the zoo because he thought she was a child. She was nearly eighteen, but she said nothing. They drove on in silence. There was so much she wanted to tell Sammy, but she just couldn't make herself talk. Her mouth was dry, and she couldn't collect her thoughts.

"A penny for your thoughts," Sammy said, turning to look at her as if he was trying to read her mind. They were sitting so close Nadia felt the heat of his body melting her fears.

"Please keep your eyes on the road. You don't want to hit somebody.

My parents would have to pick me up from the police station," she laughed nervously.

"Good point. Can't afford an accident today. We have to play it safe."

"I just don't like to sneak out behind my parent's back to meet you.

I hate lying to them." Nadia managed to say.

"Relax, don't make such a big deal out of nothing. Don't worry," he said, shrugging his shoulders. "How else are we going to meet? We're only going to the zoo." He put his hand on her thigh. Nadia shivered and felt scared of this stranger beside her who was so self-confident. She was so afraid of him. Am I making a big deal out of a simple date? Am I a coward who is afraid of taking a risk? If only my mother were closer to me, I could have talked to her about Sammy and asked her advice.

Nadia wanted to jump out of the car and return to the safety of her school, but she stayed where she was because she also wanted so very much to be near Sammy. She wanted to smell him, to touch him, to hear his voice. It was a new feeling that overwhelmed her and made her finally relax. Three months had passed since they were alone together for a few minutes at the club, and he had kissed her for the first time.

It seemed like it had happened three years ago. She had kissed him as Melanie had taught her to do. She had opened her mouth and stuck her tongue in his mouth. Sammy had seemed startled.

"Wow, where did you learn to kiss like that?" he had asked her.

"I learned from a magazine," she had said. She couldn't keep from blushing nor stop her heart from thumping so hard, but she could see that Sammy doubted her story.

Nadia had shrugged her shoulders. Let him think she was an expert.

He must have kissed a thousand girls, but he still expects me to be a virgin. Let him suffer a little.

Today, she was terrified that some relative or friend of the family might see them. She wished that she lived in America where boys came to the door to meet your parents and asked their permission to take you out on a date. She had seen it in the movies and found it a very civilized custom. In Egypt, it was unthinkable to go out alone with a boy. It was risky and forbidden. Even engaged couples had to be chaperoned.

In spite of all the taboos, the Tea Island at the Cairo Zoo was a favorite meeting place for lovers. On the island there was a picturesque open-air café surrounded by water and weeping willows. The water was full of fish and, although it was forbidden, many people threw in lines and caught fat carps for their dinner. The children fed bread to the majestic swans and the colorful ducks that floated around. You could hear the barking of the seals from their nearby island and glimpse the peacocks parading on the lawns across from the water. Nadia was very familiar with the place; she went with her family very often to have lunch there on Sundays. Today it looked different. It was a magic island where love bloomed and memories were created.

She kept telling herself that she wasn't doing anything wrong; she tried to forget her fears and enjoy the lovely day. She felt proud that they walked hand in hand like lovers; the sun was shining, and the pebble mosaics glowed like polished jewels under their feet.

Nadia tried to act grown up, but she knew she looked like a runaway child in her school uniform and scuffed shoes. What would her mother say about this escapade? Poor Mama, she wouldn't have an opinion of her own; she would run to her husband in a panic.

Nadia's father decided everything in the household, from the most trivial to the most important. He chose what they would eat every day, where they would go for the holidays, which movies they could see, what his wife wore, and who could be invited to the house. Every night he would review the household accounts and question every entry in the little black book where his wife had carefully recorded all her expenses.

Nadia hated her mother's servility and refused to become a slave like her mother when she got married. Sammy wasn't like that, she assured herself. He wasn't old-fashioned like her father.

"I can't believe that you're nearly eighteen. You sure look older," he said. "I'm twenty-three. That makes me five years older than you," he said.

"That's not a big difference. My father is fifteen years older than my mother," said Nadia admiring his curly hair and hazel eyes.

"No, it's not a big difference. We are perfect for each other. Will you wait for me? I still have one more year of medical school."

"Of course I'll wait for you." She loved his perfume; she would like to ask what it was, but was too shy.

"You never know. I bet the suitors are already knocking on your door. You are the prettiest girl in Cairo, no, in all of Egypt."

"Don't exaggerate. Anyway, I still have to finish high school and my mother is waiting for a Royal Prince. I'll wait for you."

After they finished their ice cream, Sammy got up.

"Let's go, I have to take you back to school."

"Is it already time?" she said and blushed deeply.

"Yes my princess, but we'll meet again soon. Yes?" he laughed.

On their way back, they noticed a strange commotion on Adly Street as they drove through. They had to slow down as they came upon a large group of young men walking down the street chanting, "Down with the British! Down with the oppressors! Egypt for the Egyptians!"

The demonstration seemed peaceful enough, but soon, four lorries came up out of nowhere and disgorged soldiers armed with sticks.

"What is happening? Do you think it's safe?" she said.

"I think so. The police seem to be calm; these demonstrations are never . . ." the sound of breaking glass interrupted Sammy. He turned to his left in time to see a Molotov cocktail being hurled into the display window of Groppi, the most luxurious coffee shop in town. Instantly flames started leaping into the street. Another explosion was heard behind them.

"The British Turf Club is on fire," cried a frenzied man running past them down the street.

"The British shot some Egyptian soldiers in Suez," someone was shouting.

"I'm afraid," Nadia said.

"Down with the occupation," yelled a group of young men, as they hurried towards Cinema Metro across the street.

"Maybe it was not a good day for our first date," Sammy said with a straight face.

Nadia kept quiet. They inched their way across the crowded streets of the city, which, within a few minutes, suddenly became the scene of mad rioting and looting. Panicking drivers were blowing their horns and trying to get away from the rioting mobs, and the people who were running across the traffic in alarm compounded the confusion.

Fires were suddenly everywhere. Cinema Radio in Suleiman Pasha Street was already on fire, the flames bellowing out of its arched entrance and the smoke blackening its façade. It seemed that the mob was targeting British and foreign establishments in the center of town.

They watched as the mob started breaking shop windows and looting the merchandise. Nadia started crying silently; she was gripped with fear. An angry man hit the windshield of the car with a stick, screaming, "Death to the Imperialists!" The frenzied crowds were now destroying everything in sight. They were overturning cars and setting them on fire.

"Nadia, I want you to jump into the back seat and to lie face down.

Don't be afraid. It's just a precaution." As he was speaking, a stone hit the windshield and cracked it.

"Get back there, right now," he yelled at her.

She didn't know how she got there, but she found herself lying on the back seat. She was sure that they were going to die. She told herself that she must be brave like Sammy. He never spoke another word but was fiercely concentrating on driving and avoiding the angry crowds and the panicked pedestrians. She must not be afraid of dying, for after all, she would be dying with her love.

Some angry young men gathered around the car and started shaking it. They are trying to overturn it, thought Nadia. Her mouth was dry and she could smell her own sweat.

"What are you doing, my friends? This is the car of General El Hakim, fellows. He's the head of the medical services in the army. He's not your enemy," cried Sammy. He was calm, or at least he pretended to be calm.

"You're lying," said a fierce looking man who was sweating from the heat and the excitement of the moment. He stood in front of the car barring their way.

"I'll show you the papers if you like," said Sammy opening the glove compartment and rifling through some papers.

"The police might be upon us before I find it. Shouldn't you be attacking the British and their properties instead of fellow citizens?"

"Oh, let him go," yelled a bald man carrying a huge stick in his hand. "We have to get to the British Embassy. Let's go." Suddenly, like magic, the mob moved away from the car and directed its anger elsewhere.

After what seemed to be hours of struggling against the mad traffic, during which they watched Hotel Shepherd burn, and hundreds of shops and cinemas engulfed in flames, Sammy seemed to be slowly getting away from the riots. The noise of the crowds and the smell of fire were gradually receding and the car was finally cruising quietly towards the English School for Girls in Abbaseya.

"Good girl, you did very well," said Sammy as he turned on the radio. Nat King Cole was singing "Too Young to Be in Love." It was so incongruous to hear their love song in the middle of this nightmare that it seemed to be an omen.

She looked around and saw that the street in front of her school was almost empty and very calm. "It's a miracle, I thought we were going to get killed," she said.

"I was scared too. The whole of Cairo is burning. The people are angry because the British massacred sixty soldiers yesterday near the Suez Canal. They're angry that the army didn't intervene. People are sick of the corrupt government and of the King who takes his orders from the British."

"Sammy, the car is damaged."

"We're lucky; it's only a few dents and cracks," he laughed nervously, "the most important thing is that you're safe. I wish I could drive you straight home."

"Don't worry, the driver will be here to pick me up. I think we are now out of harm's way."

"It seems quiet at the moment."

"Our first date is going to go down in history," she said.

"Sweetheart, I underestimated your power. You not only set my heart on fire but torched all of Cairo."

"Do you think it's safe for us to meet again?"

"No it's definitely not safe," he laughed, "unless you want to set all of Egypt on fire."

The next day, as Sammy was walking out of his apartment building, a police car stopped and four policemen came down and arrested him.

"Hey, what are you doing? You're making a mistake." He was manacled and taken to the police station.

"Stop talking if you know what's good for you," an officer yelled at him.

At the police station Sammy was strip-searched and fingerprinted by two police officers. He tried to speak but he was slapped by a soldier and told to shut up. They made him sit in front of an obese officer who glared at him.

"What were you doing at the English school during the riots?"

"Nothing, I was dropping a friend."

"What friend? This is a girls' school. You better tell us what you were doing there or you'll regret it."

"Nothing, I swear I was dropping a friend," said Sammy. He was beginning to get worried.

"You were seen on several occasions giving money to the guard.

What were you planning to do? Did you plan to burn the place up?"

"No, certainly not. I have no idea what you're talking about."

"Maybe we can make you remember," the officer smiled and motioned to the policeman who was holding a stick. He started hitting Sammy on his shoulders and on his back. Sammy fell off his chair and covered his face as the policeman kept on hitting him indiscriminately.

"Stop, I am innocent! My father is a General in the army. I am a medical student and have nothing to do with the riots!"

"Why were you at the school?"

"I was meeting a girl. Just meeting a girl." Sammy started crying from pain and anger.

"What?" said the officer laughing. "We have a Romeo on our hands." He laughed for a while and then stopped. "Who is the girl?"

"I'll never say," said Sammy.

The policeman was going to hit him again but the officer motioned for him to stop. "Okay lover boy, call your father and have him come here to vouch for you."

"Is that necessary?"

"Yes, it's necessary. Unless you want to sleep in jail."

He finally called his father and a furious General came to pick him up. Sammy's body was black and blue but what hurt most was the feeling of being completely humiliated. He vowed to leave this country, this burning cesspool that treated innocent people like criminals.

THE YELLOW CASTLE

WHEN MISS PERKINS, THE psychology teacher, mentioned to her senior students that they would visit the Yellow Castle, they were all intrigued.

It was the only lunatic asylum in Cairo. The name was always whispered in fear. It was used to frighten children who didn't behave. It had become part of the scare vocabulary of the local people. Insanity was a fate feared worse than death, a shameful thing carefully hidden by afflicted families from their neighbors and friends. After seeing Olivia de Havilland in the *Snake Pit*, Melanie was interested to see a mental institution and schizophrenia was a subject very much on her mind.

"My father won't allow me to go," said Nadia eating her sandwich in the schoolyard. "You know how he is. He always refuses to let me go on any school outings."

Melanie was silent for a moment. "My father lets me do whatever I want. Did you ever give your father a reason not to trust you?" She asked.

"You don't understand. We are really very different." Nadia murmured. "We are Moslems and he is very traditional and finds it very confusing to deal with a young daughter in these modern times."

Melanie looked for a long time at her friend. Could she be as innocent as she appeared to be? Was it possible for an intelligent girl to reach the age of eighteen and know nothing of the world? She was right of course; they seem to be living in different worlds. Nadia sometimes got on her nerves with her inarticulate shyness but she was her best friend and even though she was immature, she was really very smart.

She felt ten years older than Nadia and yet she knew that in reality she was the one who really needed help. The doctor had prescribed medications to control her swinging moods. Her heart skipped a beat as she thought of the Yellow Castle. Was she crazy like the patients in the lunatic asylum?

"I have a headache," Melanie finally said. She hated recess. The schoolyard was full of stupid girls in their drab beige and brown uniforms. The grass was seedy and yellow as usual. The relentless sun was too strong for a fresh green lawn to thrive and the dust from the metro that passed just behind the schoolyard made it hard to breathe.

"Melanie, they don't let me look out of the balcony anymore. They stopped me from riding horses and from riding my bicycle around the block. I haven't done anything wrong, at least not yet." Nadia laughed and two dimples showed on her pretty face. "Except that now I am eighteen, and they expect me to act like a grown up lady."

"Lady my ass! You're a prisoner, a slave!"

"It's just our tradition. They mean well."

"I'm glad you approve of your prison. It will make life easier for you. If you want to call it a life."

"Don't be so mean Melanie. I feel suffocated; I detest being a woman. I threw myself on the floor when I got my period. I screamed hysterically at my mother that I didn't want to be a girl and that I wanted to be a boy. My mother pulled me off the floor and slapped me hard."

"I feel sorry for you," Melanie said.

"I was hysterical," Nadia shrugged. "By the way, you promised to give me a book about the birds and the bees. I really need to know about life."

"I won't forget."

Melanie helped Nadia put all the pressure she could on her parents to allow her to join the field trip. A phone call from the teacher finally helped to convince them and when the day arrived, she was allowed to join the class outing.

"Do you have any crazy people in your family?" Melanie asked in a low voice. The school bus was quieter than usual. Unlike on other outings, the girls were subdued and apprehensive. Melanie sat close to Nadia in the back.

She couldn't help admiring her friend's rosy skin and soft brown hair. Nadia already had an hourglass figure and developed breasts. Melanie looked down wistfully at her flat chest, her pale skin and frail body.

"I have a cousin who is strange," confided Nadia. "My mother says she is *habla* you know . . . I don't think she is cuckoo, maybe just a little retarded." They both giggled.

Melanie thought about her mother. How crazy was she?

"My father sometimes acts crazy," continued Nadia. "He was mad that the *Moussaka* was too salty and threw the dish at the head of Osta Taha, the cook. That was definitely a crazy thing to do."

"Most people are crazy some of the time," Melanie said.

The old brown school bus valiantly fought the traffic in the old Abbasseya quarter. The smog and the choking dust were particularly unpleasant on that unforgettable summer day. The smell of garbage rotting in the intense heat wafted through the open windows of the vehicle. The girls found a diversion in watching a troop of camels that were lazily sauntering in the middle of the road. The ships of the desert arrogantly ignored the honking cars and the drivers swearing at them.

The cacophony was deafening. The skinny wrinkled camel driver, who had learned patience and forbearance from his animals, looked straight ahead. He was not aware that he was blocking the traffic. He tried to brave the noises and the traffic lights, oblivious to the danger that he and his animals were incurring. It seemed as if he couldn't wait to get back to his clean quiet desert, away from this crazy city. His hawkish black eyes stared defiantly at the disgruntled motorists.

"Are you a camel or a car?" asked Melanie.

"What?"

"You're a camel if you're trying to survive in a place where you don't belong and you're a car if this is where you belong."

"I guess I'm a car but Cairo was not built for cars." They both laughed.

"You're lucky you're not a camel," said Melanie.

The private school bus was luxurious compared to the public buses, which were banged up fuming vehicles, held together miraculously.

They seemed to Melanie to be moving only by the will of Allah. People were sitting on the roofs, astride the windows, and standing on the back fenders.

"I can't even imagine myself sitting on one of those buses. I would rather ride a camel or walk ten miles than be squeezed with so many helpless folk," said Nadia.

"I often take the bus to school," said Melanie. "It is quite an education. Once it was so crowded that a man kept rubbing against me and came all over my uniform."

"What do you mean he came?" asked Nadia.

"Ask your friend Sammy he'll explain it to you. It is something men do when they are excited."

Nadia blushed. "You're really so brave to get on those buses. They seem very dangerous."

"Life is dangerous. You have to learn to survive. Besides, these overcrowded buses are frequently the best source of jokes and funny cartoons in the papers. They are a microcosm of life in the city." Was it only in Egypt, Melanie mused, that misery was the source of mirth? It was just a defensive tactic. Humor made life easier to bear.

The five-mile drive lasted thirty minutes. When the girls finally arrived at their destination, they were covered with dust and the tainted air irritated their lungs. To Melanie's surprise, the Yellow Castle, a run-down one-story building, was really yellow. A crumbling high brick wall partially hid it from the street. It was a scary building. Reality was worse than her imagination. Would she end up there one day?

The large dusty courtyard was arid and dry. There were a few dying palm trees around the edges. Instead of plants, there were some wooden benches scattered here and there. Melanie saw a few specters roaming around like lost souls. They looked strange, almost like some other species of animals. Melanie felt scared and sick. Her fear must have shown because Nadia came closer to her and held her hand. She was pale and trembling. Miss Perkins's round plain face was red as a beet. Her round glasses slipping off her tiny nose as she ordered the girls to stay close together. Her voice was squeakier than usual.

Under the gnarled and dying trees sat a few emaciated patients.

They seemed sexless and dehumanized; their demented faces would haunt Melanie for the rest of her life. The place was dirty and decaying.

It was not a hospital but a forbidding prison. Stringy arms and legs came out of the bars on the windows on the first floor.

Melanie tried to be nonchalant and brave but failed on both counts.

She found it hard to identify with the inmates as fellow humans. She felt that their presence was an evil threat to her very existence. The doctor had told her father that she might be schizophrenic. Bipolar was the new name for the malady. It was a less scary name and evoked for Melanie a vision of the two white polar bears at the zoo in Giza. Did that mean that she was crazy? Would she end up in the Yellow Castle?

The most horrifying aspect of the place was the animal sounds that came out of those living corpses. The groans and whimpers, remained engraved on the girl's memory long after the images began fading away.

Melanie was sure she heard screams. Was she imagining them? One thing she knew for sure, all the girls were as frightened as she was. She could hardly recognize Nadia; her face was so drawn and contorted with fear. Slowly, a middle-aged doctor ambled towards them from the building. He seemed oblivious to the scene around him. He introduced himself to Miss Perkins.

"Doctor Adel Amin, the resident doctor," he said in a British accent, obviously proud of himself and his position of authority.

"Very nice to meet you Doctor. I'm Dorothy Perkins and this is my eleventh-grade psychology class," she squeaked.

Dorothy Perkins was a thirty-year-old spinster. She had come to Egypt from a small town in England to spread the word of Jesus and to teach science to high school girls. The job paid very little but it offered room and board and a chance to see the world. It also offered her a chance to heal her broken heart. She was in love with Jimmy, but he married her best friend and she couldn't bear seeing them together.

Their home town was too small to contain the three of them.

"Very pleased to meet you Miss Perkins. Welcome ladies. Would you please follow me?" The man had a respectable belly, a balding head, and wore thick glasses. He sounded weary and bored.

"Please stay together, some of the patients can sometimes be violent."

The girls didn't need to be warned. Dr. Adel escorted the fifteen girls in a businesslike fashion from one section of the hospital to the other.

He talked all the time, describing the kinds of mental illnesses treated in each section and how they tried to cure them.

Melanie noticed that he was staring at her. He made her nervous.

Did he see that there was something wrong with her? She stared back at him defiantly. She hardly heard a word he said; she was shocked by the misery she saw around her and was totally mesmerized by his low droning voice. She kept on telling herself that this was a nightmare.

Nadia suddenly grabbed her, shaking her back to reality.

"Are you going to faint?" she said. "You look so pale."

"No, someone just walked on my grave," she whispered shuddering slightly. Melanie tried to reassure her friend with a faint smile. Was she really looking that pale?

The girls were invited to visit one of the patients. The young woman was lying on a bed in the women's ward. She sat up when she saw the doctor.

"How are you Aziza?" he asked her with a smirk on his face.

"Thanks be to *Allah*, I'm well," she answered blushing deeply.

"Did you sleep well?" He asked in an amused voice.

"No doctor I couldn't sleep." She bowed her head and looked at the floor.

"Why? What's the matter?" He asked in a fake solicitous voice.

"It is the same thing every night. Some guys are molesting me all night. Please Doctor, tell them to leave me alone. I'm tired, I need to sleep and they are hurting me," she said.

"Don't worry Aziza I'll send them away," he said. He then turned to the students and explained that the woman was hysterical. She imagines that she is being raped. It is all in her mind but nevertheless her pain is real. When she was examined, she displayed signs of lacerations and inflammation in her genitals, he explained smiling.

Melanie noticed that Miss Perkins was angry and embarrassed. She attempted to end the visit. "I think it's time to go. It's getting late," she said.

"Just one last thing," the doctor said. He wasn't giving in. Melanie found herself herded with the other girls to another section where they could witness a patient undergoing electric shock treatment. The new patient was a pretty young woman who was introduced to the group as Calypso.

"*Ahlan wa Sahlan*," she said, welcoming them in a soft voice. She had a strange accent that reminded Melanie of the Greek dressmaker who lived in their building.

"You're Greek aren't you?" Melanie asked her.

"I'm half-Greek, half-Egyptian," whispered Calypso.

"*Calimera*," said Melanie shaking the woman's cold hand.

"*Calimera*," she answered sweetly. She had a nice smile and seemed to be normal. She's just like me, thought Melanie.

"Come on Calypso, it's time for your session. You want to get well, don't you?" the Doctor said.

"Yes Doctor," she answered quietly. Melanie could see that Calypso worshipped her doctor; he was her Savior, her only hope. It was also painfully apparent to Melanie that this man couldn't be trusted. His dark eyes singled her out. She felt uneasy and a wave of anger swept her body and made her sick.

"Are you okay?" Nadia said.

"I'm fine." Melanie tried to reassure her friend with a faint smile.

Was she really looking that bad?

Calypso lay down obediently. She didn't struggle as they strapped her to the special bed. The whole thing seemed like a fake demonstration.

When her body convulsed, her eyes rolled up, and her mouth foamed from the electricity surging through her. Some of the girls started crying. Melanie realized that what she was witnessing was all too real.

This was not play-acting; it was a horrible reality.

Miss Perkins became very angry. She went up to Dr. Amin and almost screamed at him. "This was not necessary. You have been rather insensitive in your choice of demonstrations. I will surely report you to the head of the hospital. I think you are sick. You're not a doctor; you should be one of the patients. Come on girls, we're leaving right now."

Melanie was so distraught; she couldn't remember how they all got back to the bus.

"If father found out about what we have seen in the name of science," said Nadia, "he would never let me go on a school trip again. He might have even taken me out of the school!"

On the way back, Melanie was unusually quiet and pensive. She kept on hearing her father saying that she was crazy like her mother.

She loved her mother even though she had run away to England and left her behind. She missed her very much. Her mother was always fun, she would chase her around the apartment and wrestle with her on the floor. She missed her hugs and her kisses. She missed her laughter and the smell of her perfume.

She couldn't forget how her mother ran naked around the house while she was doing her chores. She said that she couldn't stand the heat.

But why did she leave the windows open so that all the neighbors could stare? Did they think she was crazy?

Melanie wondered if her mother had nightmares and hallucinations like she did. There was always this dwarf that followed her everywhere she went and hid under her bed at night. Was she going mad? The psychiatrist had prescribed some pills to calm her down. Was she going to end up like Calypso? She saw that Nadia was downcast but she was too tired to think of something to say to comfort her.

"Don't worry," Melanie finally told her friend, "I'm sure Calypso is going to get well soon." Nadia shook her head and gave her an unconvincing smile.

6

THE APARTMENT

I

T BECAME EASIER FOR Nadia to invent all kinds of lies in order to meet Sammy and spend the day with him. He was her dream man. Everything about him pleased her; the crooked smile on his handsome face, the feline movement of his slim body, the artistic shape of his hands, and the golden color of his skin. She was in love. As soon as she saw him Nadia would forget her scruples and her doubts.

Three months after her eighteenth birthday, she was happy to finally skip school on a Monday and meet Sammy. He took her to their favorite restaurant that was out of the way, at the edge of the Heliopolis desert.

They went up to the second floor where there were fewer customers.

The restaurant was very plain. The floors were covered with cheap white tile and the furniture consisted of rickety tables and wooden chairs with woven cane seats. The smell of lamb cooking on the charcoal grill mixed with the perfume of Arak and filled the air.

The owner of the restaurant was a fat Greek with an eagle-shaped mustache. He was always smiling and he continuously played loud Greek music on his old gramophone. Colorful posters of the Greek Islands decorated the whitewashed walls. The smell of fresh bread baking and the salmon eggs in the *Taramasalata* was enough to make Nadia hungry. The food was delicious and the white tablecloths were white and spotless.

The upper floor was a terrace, enclosed by glass panels that were forever dusty from the desert sand and were mostly cracked from the summer heat. No one seemed to mind or notice. It was now winter and the sun filtered through the glass and warmed them as they were having lunch. At noon, the call to prayer from the neighborhood mosque drowned the Greek music for a minute or two.

Nadia watched Sammy as he ordered cold beer, stuffed grape leaves, yogurt cucumber salad, and kebab. He never asked her what she wanted to eat, but she didn't mind his taking charge. She didn't think it was an issue; her father always ordered for the whole family whenever they went out and she supposed that it was the man's job to order the food.

Anyway, it relieved her from ordering the wrong thing; she worried that Sammy might not have enough money. In spite of her school uniform, her white socks, flat moccasins, and her eighteen years, she felt grown up because she was having lunch in a restaurant with a man.

Nadia enjoyed the stories he told her. Her favorites were the stories about his childhood in London before the war. He had gone to a mixed school near Wimbledon where he had learned to tease the girls and to tap dance. He got up, and when the rhythm was right, showed her how well he tap-danced. The clients around them clapped in approval.

Sammy loved to perform.

He told her about the Beanie comics that he enjoyed reading and about the pedal boats on the river in Hyde Park.

"I'll never forget walking with my mother one Sunday afternoon, on a quiet street near where we lived in Wimbledon," he said, "and hearing the heavenly music of Mozart coming out of one of the houses. A string quartet was practicing in the living room, and we could see them from the street. That was the most civilized thing I had ever encountered."

He stopped talking to take a sip of cold beer and to munch on a crisp French fry.

"That must have been just magic," she said.

"You're magic." Sammy told her, and she felt her face turn red. She hated when she blushed but she couldn't help it.

"You know what I would like to do, my sweet? I would like very much to undress you slowly, kiss every part of your body and caress you. Then I would like to make love to you slowly and for a long time."

Nadia felt the blood rush to her groin and a strange feeling came over her.

"Sammy, stop. You're impossible," she giggled with embarrassment.

Melanie, her best friend, had just explained to her about sexual intercourse, and she had found the idea so frightening that she had stayed in bed shaking with fever for a day.

"Sex is just a way of expressing love," he said.

"Don't start again, please, Sammy. You know very well we can't."

Nadia was trying to imagine how his penis became hard and expanded.

Melanie must have given her some wrong information. It was all too disgusting for words. Nadia couldn't believe that her parents were doing it. It all sounded so ridiculous and scary.

"Many girls have sex before they're married, and then they go to a doctor to fix them up before the wedding. No one is the wiser."

"What do you mean fix them up?" Nadia tried not to show her dismay.

"Well, the doctor stitches an animal membrane on the hymen so the man thinks that he married a virgin. Sometimes the girl would stuff a chicken liver in her vagina so it looks as if she was bleeding from a broken hymen."

Nadia was nauseated; she hoped she wasn't going to get sick. Her face had blanched.

"I guess I know nothing about life." She didn't like how he always turned the conversation to sex.

"I can teach you." Sammy was eating the kebobs with relish and dipping the pita bread in the *Tehina* and *Baba Ghanouj*.

"I thought you wanted to marry me," said Nadia

"You know I do. I'm crazy about you. I'm going to go to your father, after I pass this year's exams, and ask for your hand."

"You're still a college student. He'll never accept and then he will suspect that we know each other."

"Just leave it to me; I can always fix things," he said smiling.

Nadia felt happy and warm, for a brief moment. The time always passed so fast during these mornings they were together. He would drop her at the corner of her street, and she would walk home. She had convinced her parents-with the help of a girl who lived close by-that she would be dropped off by the girl's parents on certain days when they both stayed after school for their piano lessons.

It seemed to Nadia that once she started lying there was no end to the web of deceit she was forced to weave. Her frequent absences were beginning to be noticed in school, and her grades had begun to suffer.

She was tormented with feelings of guilt and wracked with the fear of being discovered. She started losing weight and becoming silent and morose.

Whenever Nadia told Sammy that she had to stop seeing him, he threatened to do something crazy, like go to the principal, pretend that he was her brother, and drag her out of school.

She lived in torment between the fear of being discovered and the dread of losing him. She spent most of her time at home locked up in her room pretending to be studying. She was madly in love with him and she was too weak to resist his will. They continued to meet at least once a week.

The next week, he took her to a shabby unpainted building and told her that they would spend the day in the apartment of a friend, and would order some food to be delivered. This way they could be alone without worrying about being seen together.

"Sammy, how can you do this? I don't want to go up to an apartment with you. Please."

Sammy was already walking ahead of her into the building. She didn't want to go up with him, but she also knew that she had no choice.

He had made plans, and she knew how stubborn he could be. Suddenly she realized that she was at his mercy. If they ever got married, he'd be as controlling as her father, and she'd be just like her mother—a fearful and obedient wife. She had no money on her for a taxi, and more importantly, she didn't know where to go. It was too late to go back to school and too early to go home. What would she tell her mother?

She followed him reluctantly into the building and was embarrassed by the stares of the porter. The man was glaring at her with suspicion and animosity. Nadia felt his looks on her body, undressing her. Nobody had ever looked at her this way before. Her face turned red with shame and anger. She was furious at Sammy; how dare he drag her into such a degrading situation? Had he no concern for her feelings? How insensitive was he, really? At that moment all her instincts warned her against him.

Sammy took the man aside and talked to him in a whisper, giving him some money. Nadia wanted to die. Would he expose her to such suspicious hungry looks if he really loved her?

She told herself, leave, and leave right now but she waited silently for him to open the door to apartment number 32. There was no name on the door; the rusty metal frame that should have contained a plaque with the tenant's name was blank. The smell of rarefied butter, onion, and garlic that escaped from behind the badly fitted apartment doors made her sick. She shivered in the cold corridor.

"Please take me home," she pleaded with Sammy, knowing that she was wasting her breath.

"Don't be so frightened; I'm not going to eat you up." Sammy was joking, but she noticed that his hands were shaking slightly as he tried to open the stubborn door.

When they finally entered the apartment, it was very cold and had a musty smell—just like a cellar.

An old torn up sofa that was sagging with age and neglect occupied most of what was supposed to be the living room. One armchair sat apologetically on the side, looking like a bloated creature whose entrails were bursting out. She looked at Sammy, who was at the moment trying to avoid her eyes. What's on his mind? Why's he bringing me to this disgusting place? The coffee table in front of the sofa had been cleaned with a dirty rag. The marks of the wet rag mingled with the desert dust decorated the glass top with well-meaning swirls.

Sammy was busy opening a window that looked out on the backstairs of the building. The filthy metal spiral staircase served as a fire escape and back entrance for the servants. It was depressing to think that, no matter how poor you were, you could always find someone even poorer willing to serve you. The kitchen was tiny and smelled of disinfectant instead of food. A huge cockroach scuttled across the sink and vanished in a crack in the wall. Nadia stared at it, and all she could think of was that she wanted to be a virgin on her wedding night.

"Is this what they call a *Garçonniere*?" she asked.

"It belongs to a friend of mine."

Nadia could barely see Sammy through the angry tears that were blurring her sight.

"Don't be so upset, if we were living in a normal society I'd be visiting you in your home, and we'd be going out with your parent's consent. Do you think I can live for two more years without holding you in my arms? What's wrong with trying to have a little bit of privacy?"

He seemed to be making sense, but Nadia was still very confused.

At least he should have taken her to a clean apartment. This one was so sordid. Sammy went to the bedroom and opened the shutters to let the sun in. He was a sun worshipper. He sought it everywhere they went.

The sun shone into his eyes and they turned green. He smiled at her and put his arms around her and kissed her gently on the cheek.

To calm herself, Nadia looked out of the window on the street below.

A watermelon vendor had parked his cart and donkey under a huge Fichus tree. He had cut up several large melons and was shamelessly applying red powder on their greenish unripe flesh. She couldn't help smiling. Sammy came up close behind her and cupped her breasts in his hands. It felt so good to have the sun on her face and his hands on her body.

She slowly turned around, and their lips touched gently in a kiss that soon turned ravenous. He was sucking her lips and her tongue until she felt she couldn't breathe any longer. Nadia soon found herself lying on the dirty bed. The sheets were torn and had yellow stains. She struggled to get up, but Sammy held on to her.

"Let us lie down for a moment quietly; I just want to hold you close for a few minutes." He was caressing her hair, her face, and her body.

She caught her breath, wondering how he could excite her so much.

She wanted him to stop but didn't have the strength to resist. She could hear his heartbeat pounding in her ears. He kissed her neck, and she could feel his face rubbing against hers. He was undressing her slowly, and she feebly protested.

"Please, don't," she said but he shut her up with his tongue inside her mouth.

He pressed his body against hers. His hands were on her breasts and her stomach. She so much loved his face and his touch. Her feelings were so fierce; she understood for the first time the power of desire.

"Let me look at you. You're so beautiful," he told her as he gently caressed her. He slowly started taking off his clothes. She looked away, hiding her face in her arms.

"Look at me," he said and he took her hand and put it on his engorged penis.

"Sammy, I trusted you. You promised." A strong electric current went through her.

"Don't be afraid, I just want to touch you. I'm not going to hurt you," he whispered in her ear.

She felt herself losing control. She didn't know how much time had passed, but somehow he was on top of her pushing himself between her legs.

"Don't Sammy, please don't."

Suddenly he stopped moving, he was inside her. It startled them both that it was so easy. She started sweating in terror. She wasn't dead, like the girl who had died on her wedding night when she was torn open by the bridegroom's huge penis. It was a story the maid had told her in whispers.

At that moment Nadia was in shock. She couldn't feel any pain or any pleasure. The passion of the man on top of her was like a gushing river. No human force could stop it now. A wave of anger rolled over her, anger at this stranger who stole her innocence with a smile.

Her father would kill her if he found out. What if she became pregnant? What was she going to do? What if her father refused to let her marry Sammy? Would they elope together or would he take her to a doctor to fix things?

"I'm sorry; I didn't mean to go all the way." Sammy seemed to be reading her mind. "I couldn't help myself. I love you so much. It'll be all right. Don't be sad, my love, everything will be alright." He held her in his arms to calm her down. She cried inconsolably.

"Don't cry. You're going to be my wife."

Nadia continued to cry for a while and then went to the bathroom to wash herself. She looked at herself carefully in the mirror. Did she look any different? Could her mother tell by looking at her that she was no longer a virgin?

THE PROPOSAL

S AMMY SAT IN HIS room staring at the skull on his desk. He had mixed feelings about what happened with Nadia. On the one hand, he felt that now he had made sure that she was irrevocably his and had to marry him or face terrible disgrace. On the other hand, he felt guilty for taking advantage of her. He started writing her a love poem. He really loved her and wanted her to know it.

Alone all alone, my world is empty without you.
I sit in my room but I'm not at home
I'm lost. I'm nowhere. I'm wherever you are.
Alone all alone my world is empty without you.

He felt himself blushing at the doggerel he had written. She was smart and read a lot. Would she love it or would she laugh at him? He decided to mail a few words to the hotel in Lebanon where she was staying with her parents for the summer.

That evening he told his father, "I want to get married."

"What do you mean?" Dr. Hakim said laughing uneasily. "Are you joking?"

"No, I am serious. She is a very beautiful girl from a good family."

"You good-for-nothing bum! You get out of bed at noon! You never study! You barely pass your exams. What can you offer a beautiful girl from a good family?"

"I can offer her love."

50 OLFET AGRAMA

"Love! That's not a sufficient dowry these days, and I'm afraid I don't have money to pay for a real dowry. Besides, why would any man accept to give his daughter to a no-good fellow like you? Stop dreaming and get back to your books." Dr. Hakim started laughing again. He had to take off his glasses and dry them from the tears that poured out of his eyes.

Sammy glared at his father and after a pause retreated to his room.

He stared at the poster of Casablanca that hung across from his desk.

He will show him that it can be done. At the moment he was making more money than his father from the radio program he was writing and from his gambling. He would manage somehow.

Sammy passed his exams by buying the questions from a teaching assistant. He had given each one of his friends a question to research.

The day of the exam they passed the answers around with the help of the waiter who sold coffee, soft drinks and water to the students. Sammy got an A on the written exam and a D on the Oral. The important thing is that he passed. Now he had one more set of exams to pass.

Upon his return to Cairo from his summer holidays, Omar Soliman received a phone call from a stranger. "Can I meet you sir, somewhere private, at your convenience," asked Sammy in a trembling voice.

"Who is speaking and what is this about?" said Omar Soliman. He was a suspicious man.

They met at the opulent bar of the Heliopolis Palace Hotel on a late afternoon in September. Omar Soliman was wearing a dark grey suit and a *Sulka* tie. Sammy was wearing his light green linen pants, a darker green cotton shirt and an argyle cashmere pullover. They sat at a table in the bar. "What would you like to drink?" asked Omar.

"A coke please," Sammy managed to say.

After a silent pause Omar Soliman said, "Have you heard the news? I heard from reliable sources that the government has just annulled the treaty between Egypt and Britain. The British will be leaving Cairo soon for Ismailia."

"Finally, after seventy years of occupation, Egypt will be independent," commented Sammy.

"They're still in the Canal Zone," said Omar. "The Suez Canal is too important for them to let go of without a fight."

"You're right. Anyway, we're not really independent until their puppet is off the throne. I'm sure there will be protests now from the students and from the Wafd Party," said Sammy.

"Be careful what you say young man. Anyway, I don't think you are here to discuss politics. What is the reason for this meeting? What can I do for you?"

"It's about your daughter, Sir. I would like to ask you for her hand in marriage."

"How old are you young man?" asked Omar, suppressing a smile.

"I'm in the last year of medical school."

"That's good. My daughter is still in high school. Maybe you should wait a few years and ask me again."

"I'm afraid that she will be snatched away. She is too pretty and many eyes are upon her."

"And where did you see her, if I may ask?"

"At the Sporting Club. I know we have the same background and we both have a British education." Sammy blushed at his awkward reasoning. "We would make a good couple. Could you promise to reserve her for me for another year or so, till I'm ready?"

Omar laughed. "Do your parents know about your proposal?"

"Not really. I told my father but he feels that I'm not ready for marriage."

"He might be right. You are a very brave young man to approach me on your own. This is not the usual way of arranging a marriage, but all I can say is that I will not accept any offer for another year or two. She's too young."

"Thank you sir. I feel better and trust you will keep your word."

"Hey, wait a minute, I didn't promise anything," smiled Omar. He seemed amused.

All in all it was a good meeting. Sammy had another year to work things out. He went home satisfied with himself.

Sammy continued to pursue Nadia. She had refused to have sex with him. He tried to see her as often as possible but she resisted. He knew that she was scared to death of her father, and that at the same time she loved him, so he persisted. They started by seeing each other again at least once a week. People saw them together in public places and started to talk.

Omar Soliman came home one day red in the face and very angry.

"Can you imagine that Magda, my brother's wife came to the shop and scolded me? She said I had promised my daughter to General El Hakim's son and she wanted her for her own son. This is going too far. People are talking and I cannot be humiliated like this by your daughter."

"Calm down Omar we have to handle this carefully."

"What do you suggest?"

"I am not sure but the General's wife called me on the phone and wants to pay us a visit."

"Women's dirty work. What did you say?"

"I accepted of course. What could I do?"

"I could kill that girl. She's nothing but trouble. She's going to bring shame on the family."

"It's a very good family. Not what we wanted for our daughter, but they're a good family."

"I don't like to be forced into anything. Meet the woman and tell her that the girl is too young. They have to wait."

On June 12, 1952, Egypt was turned upside down. Omar Soliman came home early, he ran up the stairs laughing and yelling.

"They kicked him out. He is leaving the country!" He shouted at his startled wife. A young group of officers had staged a successful coup and the King was on his way to exile. Omar Soliman was euphoric. He hated the King and had a personal grudge against him since he had been forced to buy a very expensive gift for the royal wedding.

The revolution gave Sammy a boost. His father was a General after all and the military coup raised his status in the eyes of all Egyptians.

In the meantime another relative soon came to report that she had seen Nadia with General El Hakim's son. This was too much for Omar to take. People were talking and he decided to end this gossip and marry his daughter off. He decided to accept Sammy's proposal and announce the engagement.

Sammy and Nadia were ecstatic. There was a small party and the two families got together to celebrate. The adults were not very happy, but at least there was peace in their households. Most important of all, scandal was averted.

The next day Sammy went to visit his fiancée. He shocked his future mother-in-law by arriving wearing shorts, sandals and a tee shirt instead of a formal suit. Worst of all he had no expensive gift for the bride. Nadia's mother vent her rage upon her daughter; Nadia had brought shame on the family and will suffer the consequences for the rest of her life.

Upon leaving her that day Sammy asked Nadia for some money. "I promise to pay you in a day or two," he said smiling.

"I don't have any cash." Nadia had answered, blushing. She had spent her monthly allowance buying him a tie.

"Maybe you could lend me your watch?" he asked his trembling fiancée.

"Sure," she said after some hesitation. Her mother was right after all. She was going to suffer all her life from this unscrupulous man who had stolen her heart and soul, and now her watch.

RASS EL BARR

NADIA WOKE UP AT seven in the morning and looked out of the window of the bungalow. She took a deep breath. The sky was a transparent blue; the breeze was gentle and the sea unruffled. She put on her bathing suit and a flowery sundress. You could smell the algae that were collecting on the shore but Nadia did not see, feel, or smell anything.

She was consumed by her hatred of everything about Rass El Barr. It was so boring. There was nothing but sand and water. Nothing much happened in Rass El Barr.

Nadia heard the call of the pastry vendor who was selling *locomadis* on the beach. She smiled and rushed out the door to greet him and buy a dozen of the crunchy pastry balls, hot from a frying oil vat and dripping with syrup. She would share them with her parents for breakfast. That was the highlight of the day.

Rass El Barr was situated on the right corner of the Nile Delta, just where the eastern branch of the Nile poured into the Mediterranean.

It came into being every summer when the water of the Mediterranean receded from the north Egyptian coast, leaving at least a mile of a soft sandy beach.

In mid September the tide gradually covered the entire area and by October the area was completely under water again. The straw bungalows were rebuilt every summer on the soft sandy beach for vacationers while the Nile bank was reserved for business and nightlife.

After breakfast, Nadia and her father sat under an umbrella playing backgammon. "*Sheish beish*," he said.

She threw the dice, "*Doubara*" Nadia called out.

"That was a stupid move," he said and pushed her white markers back. She tried again and blocked his game successfully this time. She ended up winning and he closed the backgammon box with a bang. It wasn't fun playing with him. He hated to lose.

"It's my turn to play," offered her mother ready to sacrifice herself for her husband's pleasure. They played for hours on end. Nadia went into the bungalow to take a nap and dream of Sammy.

A week later while Omar Soliman was playing backgammon with his friend Hamed El Morgany on the beach, Sammy arrived carrying a battered suitcase. Nadia was delighted to see him and ran to greet him. Her father glared at her and stopped her in her tracks. The young couple shook hands demurely and giggled like children. No kissing was allowed.

"It's almost time for lunch," said Omar. "Sammy, you better check in and freshen up, we'll wait for you."

Sammy soon returned to their beach umbrella with a dejected look on his face. "There are no rooms in the hotel," he said and then burst out laughing.

"You mean you didn't reserve a room?" cried out Omar Soliman.

"I didn't think it necessary," said Sammy shrugging.

"He didn't think it necessary," repeated Omar shaking his head and looking at Nadia and Hamed.

She had wanted so much for Sammy to give a good impression but he was always embarrassing her. Her father was muttering angrily under his breath to Hamed, "This is a bad sign. He's irresponsible."

Hamed got his obese body out of his folding chair with difficulty and said in a scratchy voice: "There are two beds in my room. Sammy and I can share the room."

"No, that would be too much of an imposition," said Omar.

"Nonsense Omar. I insist and I swear by Allah the almighty and on the heads of my children that you will not refuse my offer."

"Don't swear man, may God forgive you," said Omar.

"Thank you Hamed Bey. You honor me," answered Sammy.

Everyone breathed easier and they soon went into the hotel to enjoy a huge lunch of fried fish, salad and rice.

After lunch Nadia and Sammy sat together in the lobby of the hotel.

"I have a gift for you," he said. He took out of his pocket the gold watch that he had borrowed from her and slipped it on her wrist.

"That's really a surprise," she said, blushing.

"Why? What did you think I would do, sell your watch?" he said smiling. "Don't you trust me?"

"Of course I do," she lied. "I just forgot about it," she said blushing more deeply.

"I also have an official present. This is my favorite perfume, Femme De Rochas," he said, giving her a large nicely wrapped box.

"Thank you so much. It's also my mother's favorite," she said.

"Maybe I should give it to her and get you something else."

"No, she will have to change her perfume," Nadia said. "I'll tell her it's your favorite."

"Good. What would you like to drink my love?" he asked.

She looked around hoping nobody heard him. He was impossible.

"A Coke would be nice."

He called the waiter and ordered two cokes. She was shocked to see him take out of his pocket a thick wad of bills to pay for the drinks and to later give the waiter a very generous tip.

"You brought me luck, my angel," he said, noticing her surprised look. Nadia and Sammy went swimming together in the afternoon under the watchful gaze of her parents. They were coming out of the warm sea when Omar Soliman got up and called to them. "Dry yourselves and put something on; we're going to the Nile bank for a walk. It's too hot on the beach today. I can't breathe."

They all piled into a *Taftaf*, an open carriage drawn by a jeep, that bounced on the unpaved sandy road to the Nile side of the narrow resort. The banks of the Nile always took on a festive air as the sun started to set. Ice cream shops, shooting galleries, amusement stalls, fortune tellers, *falafel* vendors, and souvenir shops stood side by side crowding each other and competing for clients.

"I should have changed," said Nadia. "I feel sticky and scratchy."

"We can swim in the Nile and wash the salt away," said Sammy.

"With your permission Omar Bey, can I hire a *felucca* to take us all up the Nile?"

"I'm not so sure that it's a good idea," said Omar. Always a no no man, thought Nadia.

"It's such a magical sunset. Please Omar Bey, let us get a *felluca*."

"Okay, maybe it would be fun." To Nadia's surprise, her father agreed.

They all got into a *felucca* and sailed up the river Nile for the half hour trip. The sun was setting and shed a golden color on the river and on the white sails of the boat that had not changed shape since Pharaonic times. They could hear the people in the other *feluccas* singing, accompanied by drums, flutes, and the primitive string instrument called the *Rababa*.

The Nile poured, at this point, into the Mediterranean and the sweet silted water of the river mingled with the salty seawater. The mouth of the river was a favorite place for dolphins to cavort and many youths loved to swim with them.

"Nadia, let's go swimming with the dolphins," said Sammy as he swiftly took off his shorts and Tee shirt and dove off the side of the boat into the water. It was getting dark. Nadia was fascinated by the sleek animals that were jumping around the swimmers in the wake of the boats.

"It's too cold and too dangerous," said Omar solemnly. Nadia imagined that he was hoping that the dolphins would devour Sammy.

He really didn't like Sammy or trust him.

"Excuse me master," said the boatman, "I beg your pardon but there is no danger from the dolphins; they're known to rescue drowning men by pushing them to safety."

"He's crazy swimming in the Nile. It's not safe!" announced Omar, scowling.

The sand from the morning swim was still itching between her legs.

Nadia decided, on the spur of the moment, to take off her sundress and jump in after Sammy. It was a supreme act of defiance. The water was warm, and though the dolphins filled her heart with terror, they were beautiful and graceful. She was afraid of their sensual slippery movements, they were creatures of the deep, and Sammy swimming among them, seemed to belong to their species.

She swam towards Sammy and could suddenly feel his sleek body slither against hers, and his hard erection jabbed into her thigh. She was aroused, and it made her panic and swim back quickly toward the boat.

"Get back onto the boat, right now," her father was screaming at them. They both obeyed reluctantly. Nadia heard the fear in her father's voice. He must have sensed the sexuality of the naked human bodies and the sleek bodies of the dolphins touching in the warm water. It was a very physical experience, almost as sensual as a sexual embrace. Nadia felt at that moment that by challenging her father, she had tied her fate forever with Sammy. She had a premonition that Sammy would always be plunging into unknown waters, and that she would have to learn to swim with him, or else drown.

On their way back to the hotel, Omar Soliman was silent and grim.

The fellow passengers in the *Taftaf* were noisy and made it easy for Sammy to whisper to her, "Meet me on the beach after they go to sleep."

"You're crazy," she told him, "the walls are made of straw. They can hear me breathing."

Sammy insisted. All the way home he kept on urging her. "I came all this way to be with you, please don't let me wait for you in vain."

"Why can't we just enjoy each other's company without subterfuge?"

All he wanted was sex. She felt trapped because she knew that in the end she would do whatever he asked her to do.

"If you don't come out to meet me," he threatened, "I'll come to your door and drag you out. I don't care if I wake up your parents."

"My father will kill you," she laughed nervously. She wanted to be with him so much, but sex before marriage was such a taboo that the thought of being discovered made her shudder. Sammy thought she was shivering from the cold. He took off his t-shirt and put it around her shoulders.

That night, around midnight, she slowly opened the door of the bungalow and snuck out. It was very dark on the beach, but she could see him sitting on a small overturned fishing boat. She ran towards him and they embraced in a long passionate kiss. All Nadia could think of was that her parents could be watching from their room.

"Let's go for a walk," said Sammy dragging her away. They walked around the hotel to a sandy alley in the back. An armed security guard was patrolling the area and gave them a stern look. It was against the law to kiss in public; they could be arrested for immoral behavior. Nadia could feel herself blushing. The few lights that had been left burning in some of the bungalows dimly lit the alleys. When they reached a dark corner Sammy grabbed her and put his hand under her skirt.

"Not here, please. Not in the street," Nadia told him.

"It's so dark," said Sammy, "No one will see us."

"Please Sammy, the guard."

"I want to make love to you. It's been such a long time. I can't stand it any longer."

"Please Sammy, I'm afraid. Someone will see us."

"You never want to make love," Sammy told her. "You're frigid.

Other girls enjoy sex, but you're always afraid."

Nadia was mortified; she couldn't bear the idea of his making love with someone else. Her jealousy and doubts turned to anger. She pushed him away.

"Leave me alone," she told him. "I don't want to make love in the street."

"If you don't let me, I'll force you. I swear I'll hit you. Is that what you want?" Sammy told her.

"Hit me then," she said her tears pouring down her face. "Hit me, I don't want to make love like a dog in an alley."

He slapped her violently on both sides of her face. She was stunned.

It wasn't only the physical pain, but the feeling of shame and anger were more than she could bear. She continued to cry but did not resist any more. She felt as if she were leaving her body and watching the scene as a spectator. She dimly realized that he had dragged her against the wall of a bungalow and was kissing and pawing at her. He was murmuring in her ear but she couldn't hear. He touched her breasts but she couldn't feel anything. She didn't resist when he took off her underpants. He put his hand between her legs and fondled her harshly. She was numb.

He maneuvered her like a rag doll against the wall and thrust himself inside her. As he was thrusting in and out of her body, she looked up and saw the stars shining brightly.

Nadia remembered herself as a child, walking beside her mother in a dark and empty street. It was during the war. On a garden fence, she saw a British soldier embracing a woman. The woman was sitting on a ledge in a strange way and the soldier's body was moving back and forth.

"What are they doing?" Nadia had asked her mother.

"Nothing," her mother had said, "Don't look."

Nadia was scared and had stared straight ahead and quickened her step. "Is he hurting her?" she had asked.

"No, I told you not to look," her mother had grabbed her roughly turning her away. They had continued to walk rapidly towards their house.

Sammy was slumping over her-finally quiet. His semen was running down her legs. She felt shaky and nauseated.

"Forgive me, darling, I couldn't control myself. You excite me so much I lost my head. Please, forgive me."

"You hit me," she said softly.

"But you asked me to hit you," he said laughing.

The house was silent and cool. Melanie had closed the heavy wooden shutters and the thick walls had kept the heat at bay. Melanie sat alone in her tiny room reading her notes on shorthand symbols. She couldn't concentrate on the work and her mind kept wandering. Suddenly, she was sure she heard voices coming from the kitchen, but when she went to check, there was no one. Her father, with his new wife and baby daughter, had all gone to the *Balteem* resort for a vacation by the sea.

She had stayed behind, in the scorching heat of August, to finish her typing and shorthand course.

Melanie went back to her room and to her notes. She knew he was there, in the corner, sniggering at her and drooling out of his twisted mouth. *Magzoub* was her faithful friend who, unlike the others, never left her alone.

"Leave me alone, you ugly monster. I can take care of myself," she told the dwarf.

"Hee, hee, I'm not so sure," he had answered very distinctly, but when she turned around he was gone.

"Leave me alone!" she almost screamed. She heard bells ringing and felt that she must be going mad. Suddenly she realized it was the phone ringing.

Nadia was calling from Rass El Barr. "Go away," she hissed at the dwarf who was lurking behind the curtain.

"Who are you talking to?" asked Nadia.

"Just an old imaginary friend," she answered.

"You're crazy Melanie. Can we talk? I'm in a public phone booth."

"What's wrong?" Melanie sensed immediately that Nadia was troubled.

"Nothing, I just wanted to hear your voice," Nadia said, and broke down crying. Sammy was with her in Rass El Barr and Melanie knew that something must have gone wrong.

"What did he do now?" she asked.

After a moment of whimpering Nadia said, "He hit me Melanie. He forced me to make love in the street."

"Jesus Nadia, tell your mother. You must try to control him. Bloody Hell, break the engagement. The son of a bitch is bad news."

"Bad news," the dwarf was saying and shaking his huge head that wobbled on his very skinny neck.

"Shshshsh."

"Melanie, I can't. My mother would die. I can't tell her. You must know that I can't. I'm no longer a virgin. I have no choice. He keeps giving me hormone shots if my period is late. I have to marry him.

Please don't worry. I just had to talk to you."

"The bastard knows that he has you in his power." said Melanie.

"He has you in his power," the dwarf whispered, echoing her words.

"I guess you're right."

"What are you going to do?" Melanie asked.

"Nothing, he's my destiny. I know he loves me in spite of everything he does. He says he can't help himself, and I do love him too."

"Nadia please think about it. He can't be trusted."

"He can't be trusted," insisted the dwarf.

"I have to go now. Sorry to make you worry Melanie, I just wanted to ventilate."

"Ventilate my ass. Next time kick him in the balls," said Melanie.

"Thanks for the advice. Bye now, I have to go." The phone clicked and went dead.

Melanie wanted to attack the dwarf and beat him to death but he had disappeared. She started to cry. She saw herself falling down into a black hole of depression. She knew that once the descent started, it would be hard to pull herself back to the surface. She cried for her classmates who were forced into marriages without love or hope; she cried for Nadia who was as vulnerable as a flower without thorns; and most of all, she cried for herself because she felt abandoned and alone in the world.

She heard the voices again coming from the next room. They were getting louder and louder. She got into bed and covered her head with the sheets, trembling with fear. Magzoub, the deformed dwarf that followed her everywhere was pulling the sheets off her head. She screamed, terrified, but then there was no one pulling the sheets after all.

THE BALLOON THEATER

MELANIE SAT ALONE ENCLOSED by the walls of her tiny room reading *Northanger Abbey*. She couldn't concentrate on the book and her mind kept wandering. The house was indifferent and cold. She was sure she heard a voice and went to check, her heart beating fast, but there was no one there.

She went back to her room and her book; but the chatty house, that was always whispering, unnerved her. To distract herself, she thought about Nadia and Sammy. A year had passed; she had finished high school and was looking for a job. How on earth did these two manage to get engaged? Was Nadia stronger than she appeared to be, or was it Sammy who had charmed and cajoled the old man into giving him his only daughter? A love match! She smiled. It was an unheard of feat that would keep the community buzzing for a while.

The Berlitz Business School was close to her home and Melanie forced herself to attend the classes that started at nine in the morning.

Every day she'd climb the dingy marble steps to the second floor and enter the dusty apartment where the school was located. She was the only one from her high school who had to find a job. She felt sorry for all the poor women who, like her, had to earn a living. "Stop feeling sorry for yourself you fool. You're on your way to independence." She smiled at the teacher who did not return her smile and then sat at her desk.

The typewriters in the school were old; some of their keys were missing. Melanie started to sneeze as soon as she breathed in the dust that covered everything in the classroom. The young man sitting in front of her reminded her of Kamal, her first and only love. What was Kamal doing at the business school? She asked herself slightly disoriented. He had the same long neck and curly black hair. She stared at the back of his head until, at some point, he turned around and looked at her with Kamal's black eyes, and for a moment she was totally flustered. Of course it wasn't Kamal.

The shorthand lesson was boring. She looked down at her notebook and tried to decipher the symbols that she had made, but instead found herself back in Balteem with Kamal, making love on top of a hill covered in Narcissus blossoms.

A bell started ringing and Melanie was startled to find herself still in the classroom. The intense perfume of Narcissus was still in the air.

Her heart was beating painfully in her chest. The class was over and she had to hand in her shorthand dictation. Her classmate got up and picked up his papers. He left the class without a second glance at her.

Melanie felt abandoned all over again. Nadia didn't call her as often as she had done when they were in school together, and when she did, it was always about Sammy.

"I'm not sure that I want to go back to school," she had said. "I just want to marry Sammy and end the suspense. My father is so angry he doesn't talk to me and my mother keeps on saying that I will never live down the disgrace. What am I to do?"

"You've got to wait one more year until he graduates from medical school, so don't stay home waiting, you'll go crazy. Jesus Christ, you're already eighteen for heaven's sake."

"And you're a wise old woman of nineteen," said Nadia laughing.

"Take advantage of the two-year Junior College that the school offers. Trust me. You won't bloody regret it. I wish I could go to school too, damn it," said Melanie.

"Why don't you try to convince your father to pay for another school year?" Nadia asked her. "I can't stand going to school when you're not there."

Melanie could still hear her father screaming at her, "I've had it.

I've paid for a decent schooling and that's enough. Go get a job and earn a living."

"Please, Daddy, all my friends are continuing their education," she had pleaded.

"Why? Are they all so ugly that they can't find a husband? If you had accepted one of the suitors I brought, you would be married by now to a decent man, and wouldn't have to worry about an education, or about finding a job."

She couldn't argue with such logic! Screw him; she was going to work and make a life for herself.

That morning there was a letter in the mailbox. Melanie became excited, hoping that Kamal had finally written to her. She was surprised that it was one of those rare letters her mother sent from England once in a while. The letter was long and described in detail how she had met a wonderful man, and that she was getting married and moving to South Africa with her new husband.

Melanie had been dreaming about and planning to save money in order to join her mother at some point, but now she figured there was no place for her in the new household. Anyway, she wasn't invited. Deep in her heart, Melanie knew that she'd soon lose Nadia just as she had lost Kamal and her mother. She'd always be an outcast. She was tired of the Egyptian boys who courted her because they only thought of her as an easy lay, and so she had stopped dating. It was her destiny to be alone.

She never revealed to Nadia how desperately lonely she was, that she sometimes heard voices talking to her; or that in bed at night, she'd hold her own hand and pretend it was someone else who was comforting her; that she was a stranger in her own home rejected by her father and his young wife.

The phone rang one evening as Melanie sat in the kitchen contemplating whether she should go to a movie or go to bed with a book. It was a cool September evening; balmy, jasmine scented nights were slowly replacing the sweltering summer evenings.

"Mel, this is Labib. How are you *ya helwa*?"

He always called her pretty one. Labib was a superb dancer, a great swimmer and a fixture at every successful party. Melanie guessed that he called all the girls *ya helwa*. He thought it was original. He was studying to be a movie director and had a flair for the dramatic.

"Bibo, how are you *ya gamil*? It's been ages." If he called her pretty one, she would call him the beautiful one.

All the girls liked Labib with his wiry body, boyish smile, freckled face, and red hair. He was living proof that Egyptians came in all colors.

Most of all, they liked him for his constant flirtatiousness.

"The club is having a party next Saturday, can you come?"

"I'm sorry Bibo, I really can't make it. I have to study. I'm taking this business course."

"Mel, this isn't a date. I want to talk to you about a project I have, and I want you to meet some of my friends who are also involved in it.

You can't hide from the world just because of studies. *Yalla*, come on, be a sport. We'll have some fun."

She found herself accepting the invitation before she was aware of what she was doing.

"Good," said Labib, "I'll pick you up, *ya helwa*, at nine. Okay?"

It took a great effort for Melanie to put on her white silk dress.

She thought about the poet Nizar Qabbani's very popular song about a woman's favorite dress talking to her of her lost love. Yes, her white dress was certainly talking to her about Kamal. She couldn't bear it.

She really didn't want to go out or see anybody. She examined herself impassively in the mirror. Her hipbones were sticking out and her arms and legs looked rather skinny. She heard a laugh. Was it the dwarf? She must remember to eat. The doorbell rang; she prepared herself to face the pitying glances of her friends.

Labib and his friend Tamer had come to pick her up. They seemed excited and happy about something.

"*Ya gamil, ya gamil*, gorgeous, gorgeous," Labib exclaimed when he saw her. "You know Tamer from the club? He's crazy about you, but then I guess we all are, aren't we?" Tamer punched Labib in the shoulder.

"Stop this nonsense, Bibo. Of course I know him," she said as she shook hands with the blushing Tamer. He wasn't tall, almost her own height, but he was small-boned and graceful. He was beautiful, in a slightly effeminate way.

"Melanie, wait till I tell you the great news." Labib was excited. "I have the money to start the folklore group I've always wanted to create.

My uncle is financing it. I'll be the director and Tamer will be the lead dancer and choreographer. We're forming a group and we want you to join us."

"What do you think? You're the best dancer we know, you'll be sensational," said Tamer.

"Damn it, I can't begin to think. Slow down and tell me what it's all about," said Melanie.

"We're recruiting ten girls and ten boys who can dance. My uncle will pay the seed money, but we are almost sure the government will give us support. There are no folklore groups in Egypt and we'll fill that void. We'll hire folk musicians from the different provinces and create dances using traditional music. All the dancers will be partners of the company and share in the profits," said Labib.

These guys are dreaming, thought Melanie. "Where are you going to perform?" She tried to be practical and not to get too excited about the crazy idea.

"We're looking for a suitable place," said Tamer.

"None of us are professional dancers," she pointed out. "We need a teacher."

"We already have a great teacher, *ya helwa*. He's an old ballet dancer from Russia," Labib replied.

"Russian? What's he going to know about Egyptian music and folklore?"

"He knows about dancing and choreography. We'll have to research and invent steps," Tamer said. He was very soft spoken but his eyes were burning with excitement.

"After all, folk dances are all alike. I mean, there are similarities, and we can adapt some of our popular dances into group dances," said Tamer.

"It sounds wonderful, but I'm afraid I need to earn a living," said Melanie.

"Oh my God! Don't we all. Even Tamer here has to make a living now; they've confiscated all his land. It's just too boring. We can't afford to pay very much at the beginning," said Labib laughing, "but I'm sure the show will be a great success and soon we'll all be rich and famous. Egypt has never had a folklore group before, and the mood of the country is crying for some nationalistic expression."

"It seems like a fine idea, but I've got to get a regular job."

"That's not a problem, we can practice in the evening," said Tamer, blushing.

Melanie liked the two young men, but she vowed there would be no romance with either one of them, or with anybody else as far as she was concerned. She made a mental note to herself that she would straighten out Tamer very quickly, if he had any romantic notions.

"Why not?" she said. "Bloody hell! What is the worst that can happen? Your rich uncle will lose his shirt, but we'll have fun. Let's do it." They all laughed out loud and left the apartment to celebrate. For a moment she was very happy.

Some months later the theatre in the suburb of Gezira was lit up and decorated with multicolored lights. The public had dubbed the place the "Balloon Theatre" and the name had stuck. Melanie arrived early, feeling very apprehensive. Her heart sank as she surveyed the theater. It felt naked, something was missing. She finally realized that the nakedness was caused by the absence of a billboard. The company couldn't afford any extra expenses, and it was decided to forgo the advertisement.

She enjoyed dancing with the group and didn't feel the need to have her name in lights, but the entrance to the theatre looked positively bare.

On the one-sheet program, Tamer Nasr el Din and Soraya Salem were featured as the company stars. It took Melanie a second to remember that Soraya Salem was her own new stage name. Melanie was "too British," Labib had declared one day, and she had agreed.

"What if nobody shows up?" Melanie asked Tamer, nervously.

"*Yalla ya* baby, come on. Don't worry; they'll come. It's something new and the people are curious. They'll come, but we have to dazzle them so they'll spread the news of how fantastic we are by word of mouth."

Melanie felt a genuine affection for Tamer. He always came to pick her up from her house and drove her back after the show. He was very attentive to her every need; he offered her sandwiches and soft drinks; he brought an extra sweater in case she felt cold after the long hours of practice. Once in a while he gave her a bar of chocolate. He was the kindest and most generous man she had ever known. He danced like a demon, with all his soul, and what he lacked in technique, he made up with enthusiasm and persistence. On stage, he made her look good.

His short stature was no problem because she danced barefoot. They were a good team.

On opening night Tamer gave her a package. "I've got a present for you," he said. When she opened it she found a small rectangular medallion on a golden chain. From the color, she could tell that it was made of twenty-four carat gold. The medallion had the throne verse from the Koran inscribed on it.

"Thank you Tamer, but that's too much."

"*Ayet al Kursi* is supposed to shield its wearer from the evil eye," he told her as he fastened the chain around her neck.

"Yes, I know. It's for good luck. I actually know the verse by heart.

Thank you, I'll cherish it, always." She kissed him lightly on the cheek.

"So you know enough Arabic to read the Koran? Melanie, you're full of surprises."

"I'm Egyptian you know, even if my mother is British."

"I really admire you. I'd like to be closer to you, more than just a friend," he said.

"Please, not tonight. Let's concentrate on the show." She hated to hurt his feelings.

"You're right," said Tamer, "my timing is wrong. Let's just dance tonight, but you must know that I love you."

Melanie put two fingers on his lips and ran off to the dressing room.

The Balloon Theater was, in reality, just a large tent that had been sold to the folklore group by a bankrupt Italian circus. The circus stalls had remained the same, and the dirt ground still smelled of elephant and horse manure. Melanie knew that it was a lucky miracle for the Egyptian National Folklore Group to have a home of any kind, but she had an uneasy feeling that the whole thing would collapse at any moment, or go up in flames. The tent was so fragile and was only held together by the ingenious handiwork of the local tentmakers and the enthusiasm of the young dancers.

In the communal dressing room, Melanie started putting on her makeup. After she put on the black *kohl* around her eyes and the dark brown wig of long hair on her head, she couldn't recognize herself. All the better, she thought. No need for friends and relatives to find out that she had become a showgirl. Her father, the hypocrite, didn't care – as long as his name wasn't implicated.

Melanie snuck a look from behind the dusty red curtain and saw that the theater was filling rapidly. Nadia and Sammy were going to be there with the Solimans. If they recognized her dancing on stage, they'd stop Nadia from seeing her and she'd lose her friend forever, or at least until Nadia got married. She hoped that the heavy makeup, the peasant costume with the fake green tattoo on her chin, and the red beaded scarf on her head would disguise her real identity.

Crowds were lining up in front of the theater. It was November, but the weather was unusually warm for that time of year. Sweating men mopped their foreheads with white handkerchiefs, and makeup ran down the faces of flushed women wearing winter coats. Now and then a warm breeze would blow from the river and moved the top of the tented theatre gently.

The audience was not the regular theater audience. The informal circus-like ambience and the cheap tickets had attracted families who had never been to a regular theater before.

The vendors were calling loudly about their wares. Sandwiches and soft drinks were being sold to those who had neglected to bring a picnic basket. It was still a circus-like atmosphere and the dancers would be the performing animals, Melanie decided. Her heart was full of dread.

She saw that the front seats were filling with the more prosperous families. They greeted and waved at each other, happy to recognize that they were among the chosen few. Old portly men proudly showed off their pretty young wives, and lovely women competed silently with each other, showing off their best dresses and most glittering jewelry.

The young girls sat next to their parents hungrily watching the young men around them who were being loud and rowdy. The smell of hashish became noticeable as soon as the young men started smoking, and gradually their voices were getting louder and their jokes more vulgar.

The musicians began to file in. They sat in a semi-circle beneath the stage, and the string players started tuning their instruments. The lute player always took the center stage, and beside him sat the man playing a small laptop harp, called the *Anoun*. A violin, a guitar, and a trumpet had been added to the otherwise traditional Egyptian orchestra. The indispensable double reed *Mizmar* and the flute didn't need tuning; neither did the various drums and tambourines.

The curtain finally rose, uncovering a village scene. A group of maidens started dancing around a water fountain, filling their pots with water and carrying them gracefully on their heads. Melanie finally appeared, and the girls surrounded her as she danced a solo; then they all joined together in a group dance with their clay *ollas*, balanced precariously at an angle on their heads. The costumes were glamorous and the girls young and pretty. When the male dancers came on the stage the audience uttered a gasp of delight. The men had on colorful wide *Sherwal* pants and embroidered vests. Instead of the traditional white clothes, they were wearing bright costumes, color coordinated with the *gallabeyas* of their dancing partners.

The stage was full of movement and color, and the music consisted of well-known folk tunes. The show delighted the audience. They were dazzled by the spectacle and gave themselves up to the flow of the familiar tunes. Their eyes could hardly take in all the brilliant costumes, the clever dancers, and the scenery that shook every time the dancers stomped their feet. Tamer stole the heart of every female in the audience with his lithe body and graceful movements. Melanie danced like a true Egyptian peasant. She was the daughter of the Nile incarnate. When the first number was finished, the audience rewarded the dancers with furious applause and cheers. The warm reception and the magic response touched Melanie profoundly. When the curtain fell after the first scene, she was in a state of euphoria that lasted all evening and brought out the best in her performance.

The smell of food filled the air. The heat under the tent increased, and many people took off their wraps and put them in the cloakroom.

Someone had the ingenious idea of creating a cloakroom out of a small lion cage. The audience left their coats with the attendants and received a number to identify their garments. Later, at the end of the show, chaos erupted. The attendants had forgotten to attach the coat check number to most of the garments, and later a free-for-all ensued.

Women were crying and men were shouting, as they searched for their coats as desperately as if they were looking for a lost child. In spite of some glitches, the evening was a resounding success. Coming out of the theatre, people sang the popular tunes off-key at the top of their voices and Melanie, or rather Soraya, had suddenly become a star.

THE WEDDING

THE WEDDING DAY ARRIVED at last and the guests started turning up after dark. The stars sparkled like crystal globes in a clear sky, big and bright, as stars tended to be over the dark desert surrounding Heliopolis. A crescent moon, sharp as a knife, slashed a huge comma among the stars dotting the sky. A heavenly smell of roses and jasmine filled the air.

The garden soon began to fill with people. All the women wore sumptuous evening gowns. Those who couldn't afford an evening dress had borrowed one from a friend or a neighbor. The multicolored dresses glittered like flowers in the garden and relieved the uniformity of the black suits worn by the men. Fine jewelry came out of safes and bank vaults, to be displayed on this special occasion, on the bodies of the women.

Melanie arrived early. She felt stupid in her virginal white taffeta dress. In spite of her objections, Nadia insisted that she had to be one of the bridesmaids. She noted that all the faces, except that of Omar Soliman, father of the bride, were glowing with happiness and excitement. Omar stood at the garden gate receiving his guests with a scowl on his dark and handsome face.

As she approached, Melanie heard him talking to Maher Shawkat, his brother-in-law, in a low voice. "Nasser just discarded General Naguib like an old shoe. I told you that he was the real power behind the Revolution.

Mind my words, as soon as he gets more established he will become another tyrant."

"Omar, I think that he is fired by democratic ideals. We have an Egyptian president who is working on a new constitution. Omar, you must be rejoicing," said Maher.

"I'll rejoice when I see free elections and political parties established."

"It'll happen. Just wait and see."

"Hello Melanie," Omar Soliman saw her and turned to greet her as she approached to shake his hand. "Nadia is waiting for you. The women are still inside waiting for the president to arrive."

"*Mabrouk* Omar Bey. Congratulations," she said as she shook hands with him and Maher. She then went to shake hands with General El Hakim and the bridegroom.

"Sammy, congratulations."

"Melanie, you look lovely," he said.

"I need to talk to you for a minute."

"Sure, what's on your mind?" he said smiling.

"As if you don't know what's on my mind."

"Melanie, this is not the right time."

"Sammy, promise me that you will not force her to get an abortion."

Suddenly there was a great commotion, Nasser had arrived and Sammy ran off to meet the president. She lost her opportunity to corner him and make him promise. The guests were herded away from the gate. Melanie had to make her way quickly through the garden towards the house and bumped into a very handsome officer who was standing in her path. He was supervising the security of the place, looking around him, his green eyes searching the guests. Their eyes met for a second and her heart skipped a beat.

"I'm sorry" he said. "Can I help you in any way?"

"I'm a friend of the bride," she said. "I have to go inside and help."

"Of course. Please go ahead before the crowd blocks your way." His smile was startling. What an idiot she was, to fall for every handsome face. She couldn't help smiling back at him.

"Thank you." She murmured and hurried into the house.

"We have to wait. The president has arrived." A breathless Amina informed the women who were in the house. "Melanie, so glad that you're here. Nadia was looking for you. I don't know what's wrong with her. She's just not herself."

"Don't worry, Mamina, she's just very excited. It's her wedding day after all."

"Yes, it's such a relief," said Amina. "I was tired of keeping an eye on her," she laughed.

Melanie stood there not knowing what to say.

"She never confides in me anymore," continued Amina with tears in her eyes. "I don't know what she's thinking."

"She's a married woman now Mamina, she'll come to you for advice and guidance," said Melanie.

"Maybe, but Sammy is her whole world now." She gave Melanie a kiss on the cheek and went away to fuss around with the other guests.

Nadia saw Melanie, rushed towards her, and hugged her. She was still in a robe waiting for the seamstress to help her with the wedding dress. She dragged Melanie to a window to watch Nasser and his men.

The excitement was great. The women were clapping and ululating like school children. He was the first Egyptian president in a thousand years.

The people had good reason to celebrate.

Melanie and Patty, the seamstress, struggled, sweating nervously, until they finally zipped up the wedding dress. It was too tight; Nadia had put on some weight during the last two months. The hormone injections hadn't worked this time.

"How are you feeling, love?" Melanie asked Nadia. "You look very pale."

"Can't stop barfing, but I'm very happy."

The women crowded around the bride kissing her and touching her. "*Bism ellah, bism ellah*, in the name of God," said Auntie Basma. "*Ma shallah, yekhzy el ein*," an old woman cried.

"May the evil eye be blinded," echoed another guest.

Everyone was talking at the same time. Laila, the bridegroom's mother, was the one who dominated the scene.

"Quiet down ladies, you are making the bride nervous," she said.

They quieted down for a few minutes, but gradually the sound increased to its previous intensity. Melanie held Nadia's clammy hand.

"I think I'm going to faint," Nadia whispered.

"Don't you dare," Melanie told her firmly.

"Nadia come and stand here." Laila ordered, and Nadia obeyed.

"Now girls, stand in a line. Two by two, the shorter girls behind the bride."

"Mummy, did the wedding cake arrive?" asked Nadia.

"Don't worry," answered her mother-in-law. "I called them five times. It's on its way."

"The caterer sent a birthday cake instead of a wedding cake," Nadia told Melanie. "Is that supposed to be an omen?"

"Yeah. It's a bloody omen. It means this is the first day of your blooming life."

"Bloody? Blooming? It'll show up. Don't worry," said Amina.

Melanie could see that Mamina was flustered. Laila, the new mother-in-law, stood sparkling like a lit up chandelier, with her silver damask dress reflecting the lights. She took control of the situation and gave orders to both the hostess and her guests in a loud, aggressive voice.

"No, young lady, you are too tall. You stand in the back." She yelled at one of Nadia's classmates, pulled by her arm and almost dragged her to the back. Amina looked worried.

"I'm afraid that Nadia is going to have a hard time with her mother–in-law," she whispered to Melanie.

"Don't worry Mamina, Sammy will defend Nadia against the dragon."

"Do you really think so?"

"I'm sure of it. He won't let anyone or anything bother his beautiful bride, not even his mother."

Mamina smiled her sweet innocent smile and seemed to be reassured. Melanie wished that she really trusted Sammy to protect Nadia from harm. The bastard, she wanted to kill him. He should have used a condom.

A young girl held a green branch over Nadia's head. Laila explained to the blushing Nadia that the branch was a symbol of fertility and would ensure her having many children. The bridesmaids giggled and laughed. Melanie smirked at the irony of the situation.

Aunt Hania placed a shallow tub of water in front of the seated bride. Rosewater and mint leaves were added to the water. Nadia was told to cross over the tub seven times. She obeyed without questioning.

Melanie wondered if anybody knew what these rituals meant. There was nothing Islamic about these superstitions, they must all be Ancient Egyptian rites.

It was very difficult to keep the wedding dress from getting wet, but all the girls helped to hold up the huge skirt above the water. Some of the older women burnt incense and held it over Nadia's head while reciting words from the Koran.

A couple of old women threw salt over the heads of the gathered women to ward off the evil eye. Nadia's school friends were as delighted as children in a charade.

When the president and his retinue finally left the wedding, the ululating became intense. Sammy arrived to claim his bride and the wedding procession started. Melanie wanted to have a word with him, but it was impossible.

Nadia was trembling with excitement and happiness as she proudly took Sammy's arm. He looked at his bride with a lot of love in his eyes and Melanie felt that maybe he would take care of her after all. Many of the women watching them were moved to tears; they were not used to seeing a couple so radiantly happy. Most brides tended to be shy and sad, and their bridegrooms to be bored and indifferent. This was a love match and everybody secretly rejoiced in it, even those who pretended to be shocked.

"*Mabrouk, mabrouk*, congratulations," the women were saying. The bridal procession started moving towards the garden.

At the head of the procession strolled six musicians, playing a violin, two reed flutes, a goatskin drum, and two tambourines. Behind the musicians were four singers, singing traditional wedding songs and clacking brass finger cymbals to the beat:

> *Oh sweet and beautiful Bride*
> *You are a perfumed carnation,*
> *Oh pretty bride you are the moon*
> *and we are the stars surrounding you.*

They were ululating and clapping to punctuate the singing. Six half naked belly dancers in glittering costumes gyrated behind the singers to the rhythm of the music. In front of the bride and groom, were four small children carrying baskets of rose petals that they scattered across the path of the procession. The singing continued:

The bridegroom is handsome like the moon
And his bride is as beautiful as two moons.
May God preserve them from the evil eye"

Six tall Nubians walked on each side of the procession holding flaming torches. They were dressed in the Turkish style, with white baggy pants and white shirts. A red cummerbund cinched their waists, and over their shirts they wore a red vest embroidered with gold thread.

Four of the bridesmaids walked behind the bride and groom carrying the bride's long veil. All the other bridesmaids, including Melanie, marched behind them in a double file. They were dressed in white and carried white candles that were four feet long, and decorated with satin ribbons, silk orange blossoms, and shiny pearls. The mother of the groom and the mother of the bride walked on the outside of the procession throwing small gold coins on the guests and the bride.

The bridal procession circled the garden twice and then stopped in front of the bridal bower. The bride and bridegroom were seated on golden chairs placed on a raised platform for all the guests to see.

The bridesmaids sat all around them on pillows scattered over the red carpeted ground. The photographer was taking pictures. The musicians continued on their path until they reached their own platform and then settled themselves down in a semi-circle. The belly dancers entertained the guests with their sexy dancing and suggestive movements.

Melanie had a lump in her throat at the sight of Nadia and Sammy.

She would never have a wedding like this one. What was she thinking?

Her father would never spend the money for a big wedding, and besides, she always felt that weddings were a ridiculous tradition. She never wanted to go through this kind of charade.

"Come on, Melanie, stay beside me." Nadia said, as she held her friend's icy hand. Melanie looked pale.

"You look tired," Nadia said.

"It's been a long day." Melanie had tears in her eyes as she hugged her friend.

Soon, it was time to open the buffet. The hungry guests rushed towards the sumptuous banquet. There were roasted lambs and huge plates of rice pilaf mixed with raisins, pine nuts, and ground meat.

There were also stuffed turkeys, boneless pigeons, roasted chicken, fried fish, grilled shrimp, and all the vegetables in season cooked in a variety of ways. The smell of mint, cumin and coriander drifted heavily in the air, mixing with the scent of fried onions and garlic. People piled heaps of food on their plates and started to gorge themselves as if they had never seen food before. Some of the men had hidden flasks of whisky in their pockets and started taking furtive sips. They were starting to get slightly drunk.

"I'm sure it was more fun watching the procession than being in it,"

the handsome officer was suddenly beside Melanie. He was looking at her with a fiery intensity.

"Yes, you might be right," Melanie said wistfully.

"Can I get you something to eat?"

"Yes, please, I don't think I have the courage to enter into the fray," she told him laughing.

"My name is Hassan," he said, as he handed her a plate with some lamb and rice.

"I'm Melanie," she said.

"I thought you were Soraya Salem, the famous folklore dancer."

"Oh, so you recognize me? That's my theatrical name. Melanie was too British," she said.

"Now you occupy the hearts of the Egyptians just as your countrymen occupied their lands."

"I'm only half British. My father is Egyptian."

"If I ever get married I would never have such an elaborate wedding," he said after an uncomfortable silence.

"I was thinking the same thing."

"Good, so we agree,"

She did not comment. For once she was not ready with a clever repartee.

"My mother wants me to marry my cousin, but I'm afraid of marriage," he said.

"I don't blame you. Forever is a serious commitment."

"Marriage is the duty of a good Moslem."

She looked up into his green eyes. "Damn it, he's religious," she thought.

Why was her heart thumping for this stranger? He touched the faint circular scar on her right cheek. "How did you get this scar?" he asked.

"A monkey kissed me," she said. "It was a pet monkey and I don't think he meant to hurt me."

"It only enhances the beauty of your face," he said.

"Thank you, you're very kind."

"There is a deep sadness in your eyes," he said.

"I'm not sad."

"Are you envious?"

She couldn't help laughing, "Yes, maybe a little."

"You're adored by thousands. If you're not a bride, it's only because you don't want to be."

She felt so alone among all those people, that she had a violent urge to fall into his arms and ask him to protect her from the world. In her sense of isolation, she yearned to hold onto this handsome stranger. She, who trusted no one, felt that she desperately wanted to trust this man.

"I must be crazy," she thought. "I will not fall in love again. I will not fall in love again," she repeated to herself. Magzoub, the dwarf had followed her to the wedding. He suddenly appeared from behind the buffet. He was laughing and pointing at Hassan. Melanie tried to ignore him and just looked into Hassan's eyes. Everything happening around them became just a backdrop; they were together and alone.

"I'll get you a drink," he said.

"Some water please."

When he left her side she decided to leave before he returned.

She walked towards the gate without saying goodbye to the bride and groom. The dwarf was following her and she could hear him laughing.

She felt too tired and confused. She hoped to escape from Hassan who she was sure would be looking for her. She asked the guard at the door to call her a taxi. As she stood waiting, it crossed her mind that she would never be able to afford a car. After the government nationalized the folklore company, she was now just a poorly paid government employee.

Suddenly, Hassan was beside her again.

"Come on, I'll give you a lift," was all he said. His red Austin Healy was right there and Melanie didn't have the strength to argue with him.

THE CLINIC

NADIA FOLLOWED SAMMY UP a dirty marble staircase. The elevator was, of course, out of order, just like all the rest of the elevators in the country.

She wondered briefly if the buildings that her father owned were as dilapidated as this one. Did he repair his elevators? She remembered hearing him complain that from the day the government fixed the rents, no one could afford to maintain their property. Since the revolution, he was continuously protesting about the new taxes and the new laws.

The smell of cooking was seeping from behind the closed apartment doors. The tenants were getting ready for dinner. The acrid smell of rarefied butter filled her nostrils and caused a new nausea attack.

Sammy wasn't looking at her; he was whistling a tune. Was he really so calm, or was he just pretending that nothing out of the ordinary was happening?

She followed this man, so familiar and yet, in many ways a stranger, up the stairs. She was afraid of him because she already knew that he could be violent. His slim body and quick movements were those of a predatory animal. The smell of *Pour Un Homme*, his favorite perfume, which she used to like, nauseated her. Did she really love him, or did she marry him because she was trapped by her fear of scandal?

The marble was worn out from the pressure of hundreds of tired feet. She had no idea that marble wore out like soap with time and use.

How many fearful women, she wondered, had walked up those steps? Did the pain in their hearts use up the marble, or were the depressions caused by the weight of their guilt?

Her own feet were of lead, and she didn't know how she moved.

Paralyzed with fear, unable to think or act on her own, she followed him silently. He must know what he was doing, she thought. A familiar sense of resignation took hold of her. Only five days ago they had celebrated their wedding. They were on their honeymoon!

"One more flight to go," said Sammy with his dazzling crooked smile that she loved so much.

"Much too expensive," her father never stopped complaining about the cost of the marriage; the wedding, the trousseau and the furniture for her apartment. Her mother never tired of repeating to her over and over again that she had sold herself cheap. She had endured the humiliation and the insults because she had no alternative.

Her neighbor Safy's fairytale marriage to King Farouk had excited the imagination of all Egyptian young girls and their mothers. Two years after their marriage, the Revolution kicked all the Royal family out of Egypt and soon after that, Safy was divorced from her King. The moral of the story was lost on Nadia's mother. Marriage for her mother was a deal, and the worth of a woman was measured by the price of her dowry, and that was that.

Sammy and Nadia finally reached the door and rang the bell. They waited for what seemed an eternity for the door to open. Nadia was looking forward to not having to lie or live a double life of subterfuge.

She was relieved to not have to meet Sammy secretly anymore. Now, she could start living away from her parents who never understood her and be an independent, married woman.

The doctor finally opened the door. He was alone in the clinic because it was after office hours. He and Sammy had been neighbors and schoolmates. Dr. Hossein had graduated from medical school two years before Sammy and was establishing his reputation as a gynecologist.

Nadia found herself alone in the tiny waiting room. The wallpaper was stained and peeling off the wall. The heat was becoming more bearable as the sun was setting. The smell of dust mingled with the odors of cooking and garbage. It was the smell of poverty, the smell of the old and aching city of Cairo, forever pushing its boundaries at the edge of the desert. It does something to your soul, this struggle to survive in an arid place.

She suddenly understood her father's point of view. He had struggled up from nothing, made money and a position for himself and now all he wanted was to be proud, to gain respect and establish himself in the eyes of the world. A beautiful daughter was an asset; through her he could finally be part of the privileged classes. He was angry with her because she had disappointed him by wanting to marry a man of her own choosing, a man from the middle class. She had forced her choice on him and he resented it. She knew that he would never forgive her.

The couch was releasing wads of stuffing from various lumps. There were garish attempts at paintings hung on the walls, in chipped golden frames. How could Dr. Hossein, so sophisticated and debonair, educated in England and a world traveler, work in such a place? Contradictions and contrasts, that was life in the Middle East. Just like her life, a fairy tale princess living in a sordid reality.

What could be more base than spending her honeymoon in this filthy place? She had talked about it briefly with Sammy. Actually, there wasn't really any discussion. He simply said that he would arrange everything, no problem. As for her, she would rather have died than have her parents find out that she was pregnant on her wedding day!

There was no hesitation on her part; she dreaded the wrath of her father far more than she dreaded the wrath of God.

The Muezzin was calling for the evening prayer. The small window in the room overlooked a narrow courtyard. She couldn't see the Mosque but it must have been quite near. The sound filled the room and vibrated against the walls.

The old sheik, who used to beat her on the legs with a stick whenever she couldn't remember the words of a Koranic verse, came to her mind.

How long ago was that? She must have been ten years old; it felt like a hundred years ago. She remembered the verses all right, but she still couldn't fathom their meaning. The Koran had a tantric rhythm that mesmerized and excited her. It was a sound that punctuated every hour of her life from the day she was born. For a while she had wanted to be a Saint, but soon discovered to her dismay that there were no Saints in Islam, and definitely no female Saints.

The wall across from the window was raw brick, colored by haphazard splashes of paint carelessly applied, and varied garbage remnants dumped out of back windows. Most of the buildings were left unfinished in the poor areas; plaster and paint were an unnecessary luxury. On the other hand, her father's villa was sumptuous and meticulously painted like any villa in the south of France.

"Come on in," said the young Doctor, "we're ready." His assistant was scrubbed and wearing a green gown. At first she couldn't recognize him, but suddenly she realized that he was Sammy. He was going to administer the anesthesia to save money, and most of all, for privacy.

There were some advantages to being a doctor, after all. He was smiling and seemed to be playing a part in a play. As the gas mask was applied to her face, she fell into a dark and peaceful abyss.

Much later, she unwillingly, gradually started to wake up on a narrow hospital bed. She had a searing pain in her stomach. She slowly opened her eyes and focused on a strange sight. Was she having a nightmare? In front of her face, a few feet away, a woman was lying on a steel bed, half naked, with her legs splayed and strapped to metal stirrups. The two young doctors were chatting and going about their business.

"She's a regular," Dr. Hossein was saying. "Because of her work, she gets regular checkups, it's good for business." They both laughed. "Once in a while, she gets careless and I take care of it."

Sammy said something that Nadia couldn't hear. It must have been a joke, for they both were sniggering like schoolboys.

"What are you doing?" She heard herself whispering. They turned around at the same time towards her, as if caught red-handed, and quickly blocked her view of the woman. Sammy didn't have to take a step to reach her, the room was so small, and he just put his hand on her head.

"What have you done to me?" She started to cry, "What have you done to our baby?" She wasn't sure if she were whispering or screaming.

The pain was so intense she thought she was dying. She hoped she was dying, for her heart was breaking with shame, with horror and regret.

"Shut up, control yourself," scolded Sammy, "Can't you see we're busy?" She shut up instantly and turned her face to the wall.

"Don't be so rough with her," Dr. Hossein said.

Later she woke up to Sammy gently slapping her face. "Wake up, it's late, it's time to go home."

She couldn't make herself wake up. All she wanted was to sleep, to escape into the peaceful void. His slapping continued steadily, drumming into her soul the painful message that she should rouse herself. In a drugged daze, she must have gotten up somehow and gone home with Sammy. She couldn't remember how or when. She found herself sleeping in their sumptuous new bed. Sammy was sitting next to her. She pretended to be asleep.

Later Nadia woke up in a sweat, crying out in pain.

"What's the matter my love?" Sammy asked in a voice husky from sleep.

"I dreamt that we were at the clinic and that the blood was gushing out of me like a fountain. My baby was born dead but he was looking at me accusingly." She started to sob out loud.

By now, Sammy was fully awake and seemed angry that his sleep was disrupted. "It's over; forget it. Come, come now, calm down and let's try to sleep a little more before we face our retribution."

It was her fate. She had to accept it and try to focus on how wondrous it was being with Sammy, starting a new life together. She had never been so close to a human being before and their intimacy was thrilling.

Soothed by the warmth of his body, she too, went back to sleep.

12

MELANIE IN LOVE

MELANIE WAS IN A taxi going home after work. She was tired and miserable. She brooded about the mess that she had made of her life.

Her health was precarious; she had just fainted during a dance rehearsal. She worried about Nadia who was recovering from an abortion.

She was briefly distracted by the magic of Cairo at night. The crowds were asleep after a hard day's work. The moon shone on the river and turned the white sails of the *feluccas* into brilliant silver sheets, fluttering sinuously in the breeze. The lights across Kasr El Nil Bridge glowed like a pearl necklace on a beautiful woman's neck. The beauty of the scene calmed her nerves.

The doctor had found out that she was very anemic and that her heartbeats were irregular. He warned her against overexerting herself and wanted her to take a holiday from her work. The thought of disrupting the dance group was unthinkable. She felt responsible for their success.

She couldn't let them down; they were her family and she owed them the only happiness she had ever known. She would never leave them.

She hoped that her health would not force her to quit the group.

She was getting a headache and could hear someone laughing softly.

She looked over at the taxi driver, but he was watching the road. She was afraid of looking next to her, for the dwarf might be lurking in the cab.

He was still following and tormenting her. She knew it was a hallucination but she couldn't get rid of him. Was she going mad?

She closed her eyes and tried to relax. The face of Hassan suddenly came to her, but she quickly dismissed him from her mind. What was she thinking? He's an impossible dream.

Melanie finally entered her sparsely furnished apartment in Garden City. She was tired and sought refuge in her private sanctuary; but instead of offering refuge, the place looked cold and threatening.

She had furnished everything in sandy beige and white. She loved the cool, uncluttered look, but tonight the white walls enclosed her like a tomb. The simple modern furniture, which she had chosen so carefully to complement her collection of modern Egyptian paintings and sculptures, now looked stark and hostile. The familiar paintings seemed alive with danger. The naïve designs of the rugs woven by young children from the village of Kerdase suddenly looked grotesque and menacing.

In the kitchen, she opened a bottle of white wine and poured herself a drink. Her hands were shaking. She took her glass and the bottle of wine to the living room, not looking at the animals and people that were beckoning to her from the walls and floors. "So what if I'm turning into a lush? Who cares?" She looked in every corner to make sure that the dwarf was not hiding anywhere. She felt him behind her pulling at her robe. She turned around abruptly. Nobody was there.

The doorbell was ringing. She stood there paralyzed with fear.

The bell rang again, and then there followed loud knocks on the door.

Finally, she forced herself to open the door. She put her arms out in front of her as if to protect herself from an intruder.

"I'm sorry, I saw you come in and was worried when there was no answer," said Hassan.

"You were following me?"

"Yes, I couldn't help myself. You've cast a spell on me." Hassan entered the apartment with decisiveness. He pulled her into his arms and kissed her passionately. His kiss was gentle but firm and Melanie responded to him.

His gentle hands caressed her tense body and she felt herself relax.

His warm lips kissed her on the face and neck. The smell of his skin excited her and the thick soft hairs on his arms sent an electric shock through her body. For a moment she was afraid that she was dreaming, but slowly she realized it was real. She had wanted him for so long and now he was here, holding her. They made love eagerly and passionately.

Their breathing synchronized like a slow dance between practiced partners.

"I've loved you from the first day I saw you," he told her.

"Why did you wait so long?"

"I thought that you belonged to another."

"Oh, you were so wrong."

"Shush, don't explain. We'll always be together now. Yes?" he asked.

He looked deeply into her eyes as if trying to read her hidden thoughts.

He had the most beautiful green eyes she had ever seen.

"Always," she said.

They finally fell asleep in each other's arms, tired but satiated, like people fasting during Ramadan, drinking water at the end of the day.

Melanie felt the warmth of his intense passion enfold her. She had nothing to hide; her soul was as naked as her body and she sensed that she had finally found her home. She was sure that she would be safe forever with Hassan. In the morning, she was afraid to open her eyes and find out that she had been hallucinating.

"Good morning beautiful," she heard his voice greeting her from somewhere in the room. She opened her eyes slowly and smiled, happy that he was near. He would chase away her demons and protect her in his strong arms. Her bed was warm and still imprinted with his body.

She felt him beside her even though he was standing across the room already showered, dressed up, and ready to leave.

"I have to go to work," he said.

"Please, don't leave me."

"I'll come back as soon as I can,"

"I can't live without you," whispered Melanie.

He sat by the bed and held her in his arms. "There's nothing I would like more than to be with you forever."

"Are you a figment of my imagination?" she said.

He hugged her with more force, "No, my love, I'm real and I'm here to stay."

"Mama, I'm home," said Hassan as he opened the door to his apartment.

"Where have you been, *ya Habibi*?" Aisha called back from the living room. "I was so worried about you." She had just finished praying and was still kneeling on the prayer rug.

"Mama, you have to stop worrying about me. I'm thirty years old.

I'm not a baby anymore." Hassan hugged his mother and helped her get up to her feet. As she removed the prayer veil from her head, he noted that her thick, soft hair was turning gray and that her movements were getting heavy.

Aisha had married Ali Hosni at sixteen, a man twenty-five years her senior. She was the daughter of a poor Turkish couple who had immigrated to Egypt. Her devout father had studied at the Islamic University of El Azhar and then had been employed in the ministry of education as a teacher. Later on, he wrote grammar books for the government primary schools.

The beautiful, green-eyed Aisha had attracted many suitors in spite of the poverty of her family. The most appealing offer came from a neighbor, an officer in the Royal guards. Her father chose him because the man was religious and traditional. Ali didn't follow the heathen European fashions of the day; he was conservative and also happened to be from a well-to-do family.

Aisha saw her husband for the first time on their wedding day, and she decided to love the stranger that fate had assigned to be her mate.

The couple led a happy, tranquil life and was delighted when their firstborn turned out to be a boy. Hassan was a healthy, beautiful baby who filled the house with laughter and joy. The happiness didn't last very long; it evaporated when, five years later, Ali died after a sudden heart attack. Aisha accepted her fate again with resignation and concentrated her love on her only son.

"My son, I pray every day that God may protect you and help you find a good woman to marry. All I want out of life is to see you settled down before I die. It's every mother's prayer."

"It seems that your wish may come true, Mama."

"Oh! Don't tell me! Is God finally going to favor me and bless me with a good wife for my only son?"

"Yes mother, I've found the woman of my dreams."

"*Bint meen*, tell me quickly, restore my heart, whose daughter is she?"

"She's the daughter of Dr. Anwar Aziz. I think he's a distant relative."

"Who? The dancer?" She beat her breast with her right hand in indignation. Aisha's face displayed undisguised horror.

"Melanie isn't a cabaret dancer mother, she's in a folklore group.

It's different. Besides my darling mother, she's going to leave the dance group after she becomes my wife. You'll love her, she's a very wonderful person."

"In spite of all the beautiful girls I introduced to you, all from good families, you had to go and fall in love with Anwar Aziz's daughter, the dancer?"

"I love her mother. Please, give me your blessing."

Tears started falling from the woman's beautiful green eyes. At the age of forty-seven, Aisha still retained traces of her previous splendor.

It saddened him that she was bitterly disappointed with his choice.

He could see that she was struggling to stop crying and to slowly gain control of herself.

"May Allah forgive me," she finally whispered, "Whatever you decide my son, I'm sure will be a wise decision. You've never disappointed me before. I just hope she's worthy of you. All I want is your happiness."

"Thank you, mother, I know that you'll love her too."

"Please, think carefully before you marry a woman who has been around. She is a dancer after all."

"Mama, you already gave me your blessing."

"Yes, I bless you with all my heart. Just promise to think about your decision a little longer."

Hassan hugged his mother tenderly; she smelled of mint and rose water. The perfume brought up sharp memories of his early childhood.

He couldn't hurt this woman who had showered him with love and sacrificed her youth for him. For a moment, he wavered in his intention to ask Melanie to marry him. He knew that she had known other men and that, as his mother said, she had been around. In addition to all that, she wasn't religious.

Would she fit in his family? Would she accept to live with his mother? Was his mother right about warning him? These questions continued to torture him, but whenever he was with Melanie he immediately dismissed his doubts. How ridiculous he was to have any misgivings. Melanie was the purest, most honest person he had ever met. She was as innocent as a child and he believed that she loved him.

Most important of all, he loved her completely.

Melanie was waiting for him in her apartment, looking radiant in her white dress that clung to her slim body. Hassan lifted her in his arms and carried her to the sofa where he deposited her gently. He was again surprised by how light she was. *She is like a bird with clipped wings.*

Now why on earth should she inspire him with such a sad image? *She's an angel, a spirit from heaven like the ones described in the Koran.*

She had cooked for him a simple dinner of roast chicken, potatoes and salad. She went into the kitchen and brought out a bottle of French champagne.

"Would you help me open it?" she asked.

"Sure, I'll open it for you."

"Oh dear, I forgot that you don't drink." She looked up at him, mortified.

"I drink sometimes, except during Ramadan. It's no big deal."

"Do you fast?"

"Yes, I do. Fasting is one of the corners of Islam." He actually blushed when he said that.

"Well, I hope you won't hold it against me that I'm not a believer."

He was silent while he filled her glass and his own with the champagne. "Of course you believe, you just don't know it," he said.

Melanie felt the blood drain from her face. She wanted everything to be perfect, but God had to come and interfere with her plans.

"Don't be so glum. Let's eat, I'm famished," said Hassan, sitting down with a serious frown on his face.

"I hope you like my cooking. I'm a terrible cook."

"It looks wonderful and it smells great." He said in an absentminded way.

"Let me serve you my love," she said as she cut the chicken and put the breast on his plate. She served him some potatoes and salad on the side and started putting food on her own plate.

"It's delicious," said Hassan, as he ate his food slowly. They ate in silence.

"Islam has given me guidance and support," he said after a few moments, "I would be lost without my faith."

"I'm sure it has, and I envy you, your certainty, and belief."

She was afraid that this might be the end of their relationship.

Hassan suddenly smiled and lifted his champagne glass.

"To your health," he said as he clicked glasses with her and took a sip of the drink.

That night they didn't make love but played chess in silence till late into the night. When Hassan left to go home, Melanie thought she would never see him again.

A week passed without him calling or coming to visit her, and when she was reduced to complete despair, he showed up one evening. He sat with her on the sofa. "Melanie, we have to get married. Will you marry me?"

Farid El Atrash was singing one of his melancholy love songs on the radio. Always about lost love and despair.

"I would love to marry you, but don't you think we should wait a few months? We hardly know each other," she said blushing.

"Why? What more do you need to know about me?"

Melanie laughed, "Why the rush? Marriage isn't important, it's just a formality."

"It's important to me, marriage is a duty in Islam."

Melanie laughed, hoping that Hassan was joking, but he was eyeing her gravely.

"What do you mean, a duty?" she sat up, her heart beating against her ribs, her smile frozen on her lips. "Are you serious?"

"I'm always serious. Would you be my wife according to the laws of Allah and his prophet?"

"I want to be your wife, but darling, I told you that I'm not religious."

Melanie felt sick. The sofa seemed to sway. She was going to lose him because of Allah and his prophet.

"The prophet married a Christian," he told her, "you're free to keep your own faith, but I would like my . . . our children to grow up as Moslems." He held her cold hands in his and smiled at her reassuringly, like a father comforting a frightened child.

"You don't understand. Even though my father is a Moslem, and technically speaking I'm a Moslem, I was brought up without religion."

Melanie was afraid to say Atheist. She didn't want to risk offending him.

"Don't look so scared, don't worry, I'm not a fanatic."

"Hassan, I love you and would do anything for you but I don't want religion to come between us."

"As you said, it's just a formality; I promise not to dictate any Moslem restrictions upon you."

"Can we just live together without a formal marriage?"

"I can't continue to live in sin. Besides, my mother would be horrified. It's important to me that we are properly married and that our children are Moslems."

"I'll have to think about that." She saw that he had not expected any objections from her.

"I don't understand you. Don't you love me?" he said.

"I love you more than anything in the world."

"If you don't believe in religion, then why not marry me in the Islamic religion just to please me."

Melanie was silent for a long while. "I'll do what you want if you agree that we won't have any children. What I want to say is, I don't think I should have any children." She was sure that she heard voices in the other room. Hassan's presence had always kept the voices away.

She clung to him in panic. He wiped the sweat from her forehead and kissed her gently.

"We don't have to worry about children now; we can discuss that later."

"Please, agree now because later I won't change my mind," she whispered.

"No children?" He asked himself. He wanted children. He had always wanted a family. His mother yearned for grandchildren, but he wanted Melanie more than he wanted anything else. Maybe he could convince her later. He took a deep breath and said, "Okay. No children."

"Great!"

The next day Hassan went to the jewelry market in the ancient Mosky Bazaar and bought a one-carat diamond ring. He had thought about Melanie and searched his soul. He had prayed and asked God for guidance and finally decided that he would depend on Allah and go ahead with the marriage.

He searched for the ring in his pocket and put it on her hand. "My angel, let's get married right away."

"I love the ring; it's just beautiful and it fits," she said.

"Just as we fit together." He kissed her passionately.

"Hassan, what about my work?" she later asked timidly. Melanie suspected that she already knew the answer, but she still had to ask.

"You'll have to resign of course, and be happy just being my wife."

"I can't just leave the group—they depend on me." She didn't tell him that she had fainted last week at the end of act two.

"We'll give them a couple of months to find a replacement. I want you just for myself."

He was strong, he was righteous, he was kind, and he loved her.

Melanie couldn't afford to lose him. She depended on him completely.

After preaching to Nadia the importance of being independent, how could she give up her freedom so quickly? The dwarf lurking in the corner made her cling to Hassan even more. She was tired and it was wonderful to abandon herself to him and damn everything else.

"I am all yours," she said.

13

EL MARG

S AMMY WAS WORRIED ABOUT money, or rather about the lack of it. He had to join the army medical corps because he couldn't find a residency at any of the hospitals. He earned very little and did not foresee a better future in the army, but he had no choice.

When he returned home late from work, tired and anxious, the sight of Nadia cheered him. She always dressed up in one of her silk dressing gowns and met him with her dimpled smile and a cold glass of wine in her hand. Sammy's heart would beat with excitement. He loved her so much. He couldn't believe she was his wife and promised himself to cherish her as long as he lived.

"*Habibty,*" he told Nadia while holding her on his lap. "My love, do you realize that I can't afford to buy you a gown like this one? It must cost more than I earn in a year."

"I have enough gowns to last me a lifetime," she laughed, "Don't worry, I'm sure you're going to be a great surgeon and make a lot of money."

"After paying the rent, we only have ten pounds left to live on." He kissed her on the neck and the cheeks.

"Ten pounds for the rest of our expenses," she murmured, cuddling up to him. Omar Soliman had forced them to live beyond their means by setting them up in a luxurious apartment and consequently, they were dependent on his financial help.

"Even with the ten extra pounds from your father, it'll never be enough," he continued. "I've decided on a new venture that will definitely increase our income."

She jumped off his lap and stood up, her brown eyes wide with expectation. "Are you going to write again for the radio like you used to?"

"Maybe, but I also have a new idea."

"What? What is it?" Nadia asked.

His face became animated. He didn't pause to take a breath but rushed on enthusiastically to explain his plan. The words cascaded out of his mouth: "I'm going to open a clinic in the village of El Marg, ten miles north of here. Last week I went to visit my father's friend, Amin Pasha, to ask him a favor. Remember their farm and their lovely villa in El Marg?"

"Oh, of course I remember."

"Well, I did some investigating and I found out that there were no medical clinics in the village. After some looking around, I found a great apartment with three rooms in a brand new apartment building. The rent is only five pounds a month. Very cheap! It would make a great clinic. What do you think?" He was absently groping for her breast.

Nadia pushed his hand away.

"Didn't you tell me once that El Marg was the center for smuggling hashish into Cairo?" she managed to ask. It was hard to put in a word when Sammy had a story to tell.

"Just listen, I'll get a friend from medical school to share with me the expenses and then, the sky's the limit! What do you think?"

"Well, it seems like a good idea. Would the army allow you to have a private practice?"

"Sure, no problem." He looked worried. "Maybe," he shrugged. "I really don't know."

"Where are we going to find the money to pay the rent, buy the medical equipment and the furniture?" Nadia asked him.

"I thought we could sell the *Sèvres* box that I gave you as a wedding gift. It's worth about two hundred pounds. Amin Pasha agreed to lend me a thousand, and if Khalil, my friend, puts up one thousand pounds, we're in business. It's a great idea, don't you think?"

"Great idea," Nadia repeated, but she wasn't so sure. She was more concerned with how to explain the disappearance of the *Sèvres* box to her parents and in-laws.

Dr. Khalil turned him down. He was doing his internship specializing in nose and throat diseases and just couldn't get involved.

In addition to that, the *Sèvres* box brought only one hundred pounds.

Sammy made a list of other things in the house that they could sell.

The sterling silver platters were valuable; he could sell them as the need arose. Sammy decided to go ahead and figure things out as he went along. He couldn't wait.

Another major obstacle came in the person of the local barber, Ahmad Ollaly. Sammy met Ollaly when he first arrived in the village of El Marg. He was a picturesque figure, tall and impressive; he carried himself like a tribal chief. A pair of handlebar moustaches dominated his handsome face. He had charisma and leadership qualities that made him well-respected and feared by the villagers. Traditionally, in his capacity as the local barber, Ollaly had performed circumcisions, lanced boils and given injections for the last fifteen years.

Sammy soon found out that the barber had chased away all the young doctors who tried to open clinics in El Marg. It was his territory and no one was going to encroach on his business. Sammy realized that Ollaly wouldn't leave him alone and that he would mount a campaign to get rid of him or at least discredit him.

He spent a few sleepless nights trying to figure out what to do.

Finally he decided that an attack would be the best defense. He was going to have a heart-to-heart talk with Ahmad Ollaly. He took himself to the barbershop for a haircut.

"*Ahlan wa sahlan*, welcome to my shop, Doctor Sammy, and to my home," Ollaly told him.

Sammy was always uneasy with people's ritual salutations. He answered the barber with a faltering, "*Ahlan beek*," hoping that he had answered correctly. Ollaly started cutting Sammy's hair with a flourish.

Sammy was very vain about his soft curly hair and realized he had taken a great risk subjecting himself to the will of his archrival. Ollaly was brandishing his scissors like a weapon, and that made Sammy very nervous. He felt a sudden desire to jump out of the chair, but restrained himself realizing it was too late. He couldn't back out now, so he ignored his fears and decided to be brave.

"Ahmad, can you tell me why all the doctors who open clinics here close down before the year is over? The peasants must be very healthy," Sammy laughed.

"People here call me Ollaly, there are so many Ahmads and Mohammeds."

"Of course, Ollaly. So tell me, what's the secret?"

"Beans, onions and lentils, that's the secret," Ollaly told him. They both laughed, the guarded laugh of two wrestlers weighing each other's strength in the arena before a fight.

"Doctor Sammy, no offense, but the fact is that the people here are so poor; they never go to a doctor until they're about to die and then it's too late. And you know what? They always blame the doctor."

"I see," said Sammy, trying to think of his next move. "How is the hair-cutting business? Can people afford to get haircuts?" Sammy was trying to gain some time; this wasn't going to be easy. He didn't know exactly how to handle the slippery Ollaly. "Business is not too good," said the smiling Ollaly. "Forgive me doctor, no offense, I get a few *malaleem* from injections and circumcisions, and those few pennies help me make ends meet."

"The bastard," thought Sammy, "he makes more money than I do." He heard himself saying, "Why don't we work together, Ollaly? You can keep all the money from the injections that I prescribe, and I'll also give you a commission on all the patients that you bring into the clinic. What do you say?"

Ollaly was beaming. "I knew the minute I saw you that you were a smart one," he said.

"Not as smart as you, my friend."

"It seems like a good idea," Ollaly told him. "But let me think about it."

Sammy rubbed his nose with his left hand and smiled his most charming smile. Son of a dog, he thought, he's playing hard to get.

A few days later, Ollaly came to the clinic before closing time.

"Let's depend on Allah," he said to Sammy. "I'm ready to go into partnership with you, doctor. May God open to us the doors of fortune."

"*In shallah*, God willing, we'll be a good team," said Sammy, his heart beating with excitement. They shook hands and the deal was done. Sammy couldn't wait to get back to Nadia with the good news.

He had turned Ollaly from a competitor into a partner. The barber of El Marg even ended up investing one thousand pounds in the new clinic.

He and Sammy were to become an unbeatable team.

"It's just like a small watch," Sammy overheard Ollaly telling one of the peasants who had brought in his sick child to the clinic. "A small watch costs more than an ordinary watch because it's smaller and its inner workings are more delicate. Your daughter is like a small watch and therefore more expensive."

Ollaly was trying to convince the man to pay ten pounds for an appendectomy for his child, instead of the usual five. Sammy smiled and marveled at the man's skill in public relations. Ollaly was going to really help make the clinic a success.

Sammy never made house calls; patients usually came to the clinic to be examined. The village was small and the distances were short. The main street was a narrow dirt road with a few shops on both sides. The fields were not far away. One day, however, Sammy was summoned to call on a patient who couldn't make it to the clinic.

He drove along the dirt road for about three miles in his Fiat.

Away from the main street with its row of shops, Sammy looked with renewed dismay at the scattered homes in the village that were made of mud bricks. They had no running water, gas or electricity. The villagers shared their primitive lodgings with their animals. It could have been a village built five thousand years earlier.

The flat countryside was green with fields of corn, beans and wheat.

The skinny cows grazing in the patches of clover definitely looked undernourished. The water mills, drawn by thin oxen, drew the water from the nearby tributary and watered fields that were crisscrossed with shallow trenches. The ancient Egyptian irrigation system was still working just as it had since the dawn of civilization.

Sammy had to lower his head to enter the dark one-room house.

An old man lay on a straw mat, in the one and only room. He was obviously suffering from great pain. Sammy examined him on the floor and realized that he had a ruptured appendix. He needed to have an emergency operation. The man had waited too long and now there was no time to waste.

"Ollaly, we need to operate on the *Haj* right now! Where's the nearest phone? I'll call the military hospital for an assistant."

"Can't we move him to the clinic instead?"

"No." Sammy looked around anxiously for an operating surface. "I need you to disinfect the top of this brick oven and to lay some clean sheets on it. Where can I find a phone?" he asked, looking at the people standing around him in the room.

"The nearest phone is in the pharmacy down the road," said a young boy of about ten. "I'll show you the way. Follow me."

"Ollaly," Sammy called after him, as he ran out of the adobe hut, "we need some boiling water and clean towels. I'll see what I can round up from the hospital."

The boy put the hem of his long *gallabeya* gown between his teeth, and took off on his bicycle with Sammy following in his car.

An hour later, a military jeep arrived, laden with instruments and medications. The operating room was ready. Soon, Sammy was performing the operation on top of the brick oven.

"Now Ollaly, hold this lantern above my hands so I can see what I'm doing. Dr. Hisham, start administering the ether."

"Keep the damn lantern away from me and put out your cigarette or we'll all be blown away," Dr. Hisham told Ollaly nervously; he was trembling.

"Depend on God, doctor," Ollaly said. "He'll protect us."

Hisham kept muttering oaths under his breath as he signaled Sammy to start.

With a gas lantern as his only source of light, and instruments and drugs borrowed from the army hospital, Sammy performed what the peasants considered a miracle. The proximity of the gas lamp to the highly inflammable ether was a risk he had to take. Luckily, there was no explosion and the operation proved to be a success.

"All this work for what?" he complained to Nadia after the grueling day at his clinic. "I don't think I'm going to get rich being a village doctor. The work is too hard, and besides, I hate taking money off the poor peasants."

"You're doing great," Nadia told him. "Just be patient, soon you'll have a regular hospital."

"You're such an optimist."

Sammy's patients in the village of El Merg considered what they owed him to be a debt of honor. They paid their bills whenever they could, and that meant waiting for the sale of a crop or for a seasonal animal market. He was sometimes paid in hashish and at other times with stuffed pigeons, fresh eggs, or a *tajin* of meat and potatoes baked in the village oven.

The hashish he rarely smoked. He had been wary of hashish ever since the day he almost ended up in the hospital when, as a joke, some of his friends slipped a large piece of hashish into his coffee. Instead of turning down the hashish, he sold it to his fellow doctors. The food he brought home to Nadia.

The couple would invariably run out of money around the middle of the month, but they never went hungry. If they didn't get food from the patients, they would drop in for a meal at one of their parents.

Whatever Sammy earned at the clinic, he spent swiftly; on a gift for Nadia, dinner in a restaurant, or at the movies. They loved to watch two movies back-to-back.

"I hate to pick the pockets of the poor people," he constantly complained to Nadia.

"We should save part of your earnings for emergencies." Nadia advised, although she herself had very little concept of handling money.

"It's not worth saving a few pounds. I'll start saving when I have thousands of pounds, maybe millions, but not before. Poor baby, you've married a poor man."

"Money isn't important," she insisted.

"You say that because you've always had it," he told her. "Have you ever worn torn socks? Do you know that the first time I took you to the apartment I had to buy new underwear?"

"We have everything we need," she said.

OUT OF THE HAREM

NADIA CAME IN FROM the kitchen proudly carrying a silver tray with two delicate teacups and a plate of her mother-in-law's raisin-hazelnut cookies. Melanie sat in the living room waiting and taking in her luxurious surroundings.

"I am so happy for you," said Nadia, very excited by the news of Melanie's engagement. She was waiting eagerly for her friend to fill her in on all the details of how and when. Melanie remained silent. Nadia poured the tea and waited patiently for her friend to talk.

"I thought you would never get married," Nadia said, laughing nervously. "You always found something wrong with every man you met. This Hassan must be a paragon."

"Yes, he is the man of my dreams and I can't live without him," said Melanie without smiling.

"Why the long face? Something's wrong?"

"No, I'm just worried about my work, trying to get my life in order."

"It seems to me like you are regretting your quick decision. When and where did you meet prince charming?"

"We met at your wedding, three months ago."

"And you kept it a secret from me?"

"Nadia, I wasn't sure. It happened so fast."

"Well, never mind, I also have a secret. Guess what?"

"You're pregnant," said Melanie smiling.

"How did you guess?"

"It's not difficult to come to the conclusion, you look radiantly pregnant."

"I'm so happy. I love children and I want Sammy's baby."

"You're so young. You're going to be a teenage mother."

"It's so exciting. There's only one problem. We don't have any money."

Melanie motioned with her arms at all the luxury surrounding her.

"What are you talking about?"

"My father insisted on furnishing this apartment for us but it's only for show. He wants to save face in front of family and friends but he will not help us pay the rent. Sammy's salary hardly feeds us."

"Maybe you could eventually get a job and help with the expenses."

"Melanie dear, it's easy for you to say. You're lucky you have such a glamorous job. What kind of a job can I do?"

"You can go to the university and get a degree and teach, or work for a newspaper, or in public relations."

"I would love to be a professor like my aunt Dora. She was the first woman in Egypt to get a Ph.D."

Melanie started laughing nervously.

"What's so funny?"

"Here's to the beginning of your career, and to the end of mine."

"What are you talking about?"

"He wants me to leave my dancing and stay home."

"I can't believe that you agreed. It is against everything you believed in."

"We're still negotiating." said Nadia.

"If you're still negotiating, then I'm sure you'll have your way."

"I wish I had as much confidence in myself as you have in me, girl.
I love you."

Following what had become a routine, Sammy and Nadia went to have lunch at Sammy's parent's every Friday. When they arrived at the Hakims, Laila embraced Nadia with cold detachment; she reserved her

loving kisses for her son. She sat beside him on the dusty sofa in the den and started fawning on him and kissing him. What a perversion! It was very embarrassing to watch and Nadia sat there blushing.

"What's the matter?" asked Laila, "Are you jealous?"

Nadia didn't answer, and Sammy got up and left the room.

"I heard that you were pregnant," she smirked.

Nadia kept silent, no congratulations? She nodded her head in agreement.

"What's the hurry?" she asked. "You'll ruin your figure quickly and he'll start looking at other women. I know my son; he'll lose interest in you. You should've waited a few years."

To Nadia's relief the other guests started arriving and Laila left her alone in order to receive her visitors. The guests gathered around Dr. El Hakim in the living room in a relaxed way. There were now ten people crowded into the cluttered room, which was less than perfectly clean and untidily chaotic. There was dust on the furniture and books on the floor. The sofa and chairs needed recovering. The material was faded and worn out around the edges. Laila was always too busy, raising money for charitable organizations and being on the board of a home for orphans. She wasn't interested in housekeeping.

When all the guests were finally seated at the dining room table, it was after two in the afternoon and everyone was starving. Laila hovered around the table. She had the irritating habit of never sitting down with her guests, but preferred to stand up, serving everyone and making sure that no one was neglected.

Muhammad, the cook, came into the dining room carrying a huge bowl of okra swimming in tomato sauce. Small pieces of lamb were floating in the sauce and the smell of cumin was already permeating the dining room. He had been working for the Hakims for twenty years and was considered one of the family. Osta Muhammad was a typical Egyptian peasant. He had a pleasant clean-shaven face, perfect teeth and a light-brown complexion.

"*Salamo Aleikom*, you have honored us," he greeted the guests. He acted as if he were the host and shook hands all around after he had set the food on the table. He moved around the house majestically in his long striped *gallabeya*. The man cleaned the house, ran errands, and helped Laila prepare the meals.

"It's a national disaster," said one of the guests. "He was one of the old school generals who was forced to retire after the revolution. This morning Shammas Pasha dropped dead as soon as he heard that his land was confiscated, and Nadir El Halawany shot himself in the head."

The rest of the guests stopped eating in shock. The old general seemed to relish the effect of his gruesome gossip.

"Mark my words," Laila said in a loud voice, "the big landowners are going to be begging in the streets. You're lucky Sammy, at least you married into a merchant family. They'll keep their money—for the time being anyway," she laughed a shrill and nervous laugh. You could see horror written on the faces of the men sitting around the table. They weren't used to having women speak out in mixed company.

Dr. El Hakim raised his eyes and looked at his wife with disapproval, but she ignored him, as was her habit.

"I just can't imagine how the peasants are going to be able to manage their affairs on their own," said Hamed, Hakim's brother. No two brothers were less alike. Hamed was a giant with a big belly and a huge chest. Dr. El Hakim was short and slim.

"How will a peasant know how to pay the taxes, where to buy the seeds, and when to sell his harvest?"

"Don't worry, Hamed," Laila said, "they'll learn very fast."

"I just hope they'll improve their life and not just spend the money on second wives," said Sammy laughing.

"You're right Sammy, men are all the same; they only think with their you-know-what, whether they're educated or ignorant," said Laila.

No one laughed. All the men around the table were secretly glad that Laila was not their wife.

There was no ceremony at the Hakims' table. Some of the guests ate with their hands, the way the peasants did in Egypt, using pieces of pita bread to eat their vegetables and dip into the rich tomato sauce.

The crude eating habits of some of the guests did not bother Nadia. She enjoyed dipping her bread into the sauce too, something that was strictly forbidden in her family. She felt a great sense of freedom and the food tasted better. The conversation around the table intrigued her; it was a sharp contrast to the silent monosyllabic meals at her parents' home. She was sitting on the right of Dr. El Hakim, and she became aware that he had aged immeasurably since she had first met him. What was left of his hair had turned gray and his eyes seemed to be bothering him more than usual. He caught her eye and smiled affectionately. He motioned to the guest sitting beside him and said to her, "We're very lucky today.

My friend Dr. Nabil Mahmoud has agreed to entertain us after lunch. Do you know that he's an excellent *Oud* player as well as a great poet?"

"I've heard you many times on the radio, Dr. Nabil," said Nadia. It was too bad that Moslem families didn't celebrate Christmas. He would have been a perfect Father Christmas.

"I think it's wonderful that you're a writer, a musician and a doctor," she told him. "Do you know that Sammy is also a writer? He writes regularly for a children's radio show and is now writing a movie script."

"That's a surprise, Sammy. I didn't know that you were a writer. You know, I believe that without words and music, life would be unbearable."

"I'm not a great poet like you, Doctor Nabil," Sammy told him, "but I do love to write, and lately I've been writing a movie script based on *Pygmalion*."

"The story of *Pygmalion* has inspired many writers. Are you being influenced by Shaw or by the Greeks?"

"I'm afraid I'm not an intellectual Dr. Nabil, but Shaw was my inspiration."

"I've a feeling you're going to have an interesting script on your hands. Good luck."

"Thank you Doctor, for your blind faith."

"We artistic doctors have to stand up for each other, don't you think?"

"I hope Sammy is half as good as you are, my friend," said Dr. El Hakim. "I'm afraid he's lost between medicine and the film business."

"Don't worry, Youssef. Your son is smart; he'll find his way."

"From your mouth to the gates of heaven," said Laila, lifting her hands in prayer.

"What about you Nadia, do you have a hobby?" Nabil asked her.

"I used to play the piano," she said, "and I love to paint, but I'm afraid I've stopped all these things since I got married."

"She's devoting herself to me now," Sammy told him laughing, "I'm her new hobby."

Nadia blushed. Dr. Nabil smiled and said, "Young love is a wondrous thing to behold."

El Hakim turned to Nadia, "I believe that love is not enough, one should also pursue his personal interests. You can't live your life through another person. Tell me honestly, my dear, aren't you bored waiting for Sammy to come home? What do you do all day?"

She would have liked to tell her father-in-law that she slept late, threw up every morning, read in bed and talked on the phone for hours, but all eyes around the table were upon her and it made her self-conscious.

"I do very little, I'm afraid." She knew that she was blushing fiercely and felt like a fool. "A friend of mine just suggested to me that I should go to college and get a degree."

"What a great idea!" said Dr. El Hakim. "Why don't you go back to school and get a degree? Great idea." He removed his thick glasses and rubbed his eyes.

"Is it possible?" she asked, surprised. Melanie was right after all.

"Can I still enroll at the university?" Her heart began to beat rapidly with expectation.

"Certainly you can, my child. Would you really like to?"

She looked questioningly at Sammy, but he seemed to be listening to the conversation going on between two ex-generals and was silently rubbing the side of his nose with his left hand. He always did that when he was upset.

"You know uncle, my father is against a college education for women."

There was sound of nervous laughter scattered around the table. She suspected that many of the men agreed with her father. El Hakim looked at her earnestly, thinking to himself that she was looking especially radiant. Her shiny, light brown hair had grown to shoulder length, her brown eyes were sparkling, and her pale skin had taken a rosy hue. He could see that she was no longer a child, but was becoming a woman.

El Hakim smiled and said in his soft voice, "I think going to college would be far better than wasting your time playing house. It is important my dear; times are changing." He glanced at his son, but Sammy was busy eating and remained silent.

"Learning would enrich your life," said Dr. Nabil, as he took a second helping of rice and piled more okra on top.

"I would love to go to college," whispered Nadia.

"Then you must!" El Hakim answered emphatically.

Nadia adored her father-in-law and would have liked to hug him right then and there. Sammy was smiling and seemed amused at this new prospect, but he continued eating and made no comment.

After lunch, Dr. Nabil played some of the old popular tunes of Sayyed Darwish on his big bellied *Oud*. The instrument was the father of the modern guitar. It reflected the earthy soul of the Egyptian peasant, a feeling that was lost in modern Egyptian music that was trying hard to mimic the West. The doctor-musician changed before her eyes from a fat old man into Pan, the God of music.

Later that night when they were back home, Nadia threw herself on the bed and asked Sammy to rub her back; she was aching and his massage felt very good. She hoped that he would leave her alone that night and let her sleep. She enjoyed making love but their intense sexual activity was getting to be exhausting. Most of the time now it was actually painful. Whenever she complained, Sammy would say, "It's supposed to hurt."

"Tell me," she asked him later, "what do you really think of the idea of my going to college?"

"I personally don't think it's such a good idea," he shrugged. "You'll soon be very busy with the baby, but if it amuses you, I'll let you do it."

"Amuse me?" She sat up straight. "I thought I might be able to work and help you with the expenses."

"How much do you think you would earn as a teacher?" He had a cynical smile on his face.

"Enough to make us not run to our parents all the time for a free meal."

"I'm sure that I'll be able to take care of you and the baby."

She looked at him silently.

"Well, I guess you should try it," he said. "At least it would give you something to do."

"Thanks for giving me permission. You're sometimes very insensitive and patronizing."

"Let's not fight." He embraced her and started kissing her passionately. "I love you so much," he said, "I'm just being selfish. I want you all to myself—but I also want you to be happy. If you really want to do it, I'll support your decision."

They made love and all the time she couldn't help worrying if it was safe for the baby. She was too embarrassed to ask her doctor and he hadn't mentioned anything about it.

Two days later on Sunday, they went to have lunch with her parents.

The house was silent and empty. Her brother was away on a school trip to Europe. There were flowers in all the vases, all the clocks were working, and there was candy and chocolates on every table around the house. The dining table was impeccably set with antique Limoges china

from France, Baccarat crystal goblets and sterling silverware. The table that could seat sixteen comfortably was set only for four.

While they were having lunch, Nadia told her parents about her new plans. Omar Soliman exploded.

"I didn't give you my daughter to prostitute her!" he yelled at Sammy.

Nadia almost jumped out of her seat in shock.

"How can you allow her to mingle with all the young men at the university?"

Morsy, the Nubian butler, came in at that moment with the appetizer and served them in stony silence, but with efficiency and grace. He was dressed in his Turkish costume of wide white pants, white shirt, embroidered vest, red cummerbund, and a red fez on his head. His face twitched slightly when Nadia smiled at him. The man was used to being invisible, so he was afraid to acknowledge her greeting.

Sammy waited until Morsy had left the room. "Education is not prostitution, sir," he said, trying to control his anger.

"I bet the next thing you're going to tell me is that you're going to put her out to work?"

"After she gets her degree she might want to work; I've no objections."

"No objections? I object strongly!" Omar Soliman banged his fist on the table. The crystal goblets vibrated and Nadia cringed. She was glad that there were only the four of them at the table.

"A woman's place is in the home with her children, not running around the streets. No good is going to come of these new ideas."

Nadia watched the discussion silently, while eating the delicate spinach soufflé. No one was bothering to ask for her opinion. In spite of being invisible, she suddenly felt nauseous.

"There are women in your own family who have had successful careers," Sammy told Omar. "One of them is a professor at the University of Cairo and another is under-secretary to the Minister of Education. They both enjoy the respect of the whole community."

Sammy continued to eat his soufflé in spite of the arguing going on.

Omar looked at him hatefully and yelled. "Respect of the community? These women you're talking about are worthless mothers and wives!"

Nadia was terrified by her father's outburst. She was afraid that he was going to choke on his food. He reminded her of the verse which said:

"and lower thy voice;
For the harshest of sounds, without doubt,
is the braying of the ass."

She remained silent and kept her eyes on her plate. After a few moments of stillness, while Omar was thinking with rage about the errant career women in his family, and Sammy was fantasizing about the best way to murder his father-in-law, Mrs. Soliman finally collected her courage and murmured, punctuating her speech with her peculiar little laugh.

"Heh, heh, now be reasonable, *mon cher*, heh, heh, heh, college never hurt anyone. This is 1955, after all." She had this short silly laugh that drove Sammy crazy.

"I'm discussing this with Sammy. I didn't ask for your opinion, woman."

Amina was deeply embarrassed; she controlled the tears that filled her eyes, looked down and kept quiet. She groped for Nadia's hand and held it in her sweaty palm. The talcum powder, which she loved to pat abundantly all over her body, was still visible on the exposed part of her heaving chest.

"Could you please tell me if you have figured out how she would be getting to the university every day?" Omar asked Sammy in a derisive tone. A vein was visibly throbbing in his forehead.

"She'll take the Metro," said Sammy.

Nadia could see that her father's anger was rising, the way foam rises to the top of a cup of Turkish coffee.

"Never!" Omar banged with his fist again on the table.

Amina stiffened up in her seat but kept quiet. Morsy came in with a lamb roast surrounded by fried potato croquettes and glazed carrots, all served on a huge sterling silver platter.

"My daughter isn't going to be dishonored by riding the tram with the working people. If she must go to the university, I'll send her the car with the driver."

Sammy took his time and helped himself generously from the platter presented to him. He was especially fond of roasted lamb. "It's very kind of you, sir," Sammy finally said, "but there is really no need. The metro runs

in a straight line from near our apartment up to the gates of the university."
Sammy was making an effort to control his anger and started pouring some
sauce on the large piece of succulent lamb on his dish. He had decided that
Omar's obnoxiousness wasn't going to spoil his appetite, but he couldn't stop
his hand from shaking and spilled some of the sauce on the spotless linen
tablecloth. "I'm sorry," he said to Amina. He scraped the sauce off with his
knife and started pouring some salt on it, in the hope of removing the spot.

"Heh, heh, don't, don't worry about it, my dear, please," she said.

She discreetly threw some salt over her shoulder to counteract the bad
luck. They continued to eat in silence.

"What kind of a man are you?" continued Omar after awhile.

"Doesn't it bother you to expose her like that?"

"What kind of man am I? I'm a man who loves your daughter and
would rather die than expose her to any harm."

"No," Omar yelled, "No daughter of mine is going on a tram like the
rakash, the common trash. Never!" He glared at Sammy. "I'll send the car."
Omar's voice was breaking up from yelling so loud.

"Well," Sammy was getting impatient, "if you insist, we'll gratefully
accept your offer, thank you, sir. That means that you bless our plan with
your approval?"

"No, you must know that the whole idea is against my better judgment
and against my wishes."

Sammy got up suddenly from his seat and stood behind his chair.

He had had enough of Omar Soliman and his arrogance. "Don't worry,
my wife will enjoy college. It won't hurt her. Come on Nadia, let's go," said
Sammy.

"Heh, heh, you haven't had your dessert yet," said Amina. "We have
créme caramel, your favorite."

"I have an urgent appointment. Come on Nadia, let's go."

Nadia stood up, kissed her mother good-bye and went over to kiss her
father. Omar didn't return her kiss. Sammy would've loved to tell Omar
Soliman to go to hell, but he held his tongue out of respect for Nadia's
feelings. The man can take his money and shove it. There was no need to
talk to him anymore, thought Sammy.

EIN SHAMS UNIVERSITY

THE COLLEGES OF EIN Shams University were scattered all over the old quarter of the same name. The classes were held in old schools and makeshift buildings. It was a good university, but had no campus of its own. The faculty of literature was located in a run-down building with small rooms and a dusty yard that surrounded the building on three sides. A few giant trees provided some shade to the students milling around between lectures.

There was no recreational area; the students sat on the fence around the school and talked while they ate their sandwiches, brought from home, wrapped in old newspapers. The library was in a dusty basement with iron bars on the windows. Very few books could be checked out and students had to sit, like penitents, on hard wooden chairs in the semi-darkness to read the books they needed. Most students couldn't afford to buy new books, so the professors made smudgy dictographs of articles and chapters for each lecture and passed them around the class.

Nadia was the only student to arrive at Ein Shams University in a car with a driver. She was embarrassed by the curious looks that met her. She talked to some of the girls but found it difficult to make any friends. She was wealthy, and most of them were poor; she was married and they were single, so it was only natural that they regarded her with suspicion.

The first lecture she attended was about Shakespeare's comedies.

The professor, an Egyptian, spoke a beautiful English acquired at Oxford University. He had a very handsome face and bright black eyes that were enlarged by his thick glasses. He was a short man, but standing on a raised platform, he looked impressive. He had a strong pleasing voice and he lectured without a pause, constantly referring to his notes.

The students were supposed to write down every word and memorize it. He deliberately spoke slowly, conscious that he was addressing a group that was not fluent in English. He interspersed his lecture with Arabic commentaries in order to clarify some points. The mixture of Arabic and English confused Nadia at the beginning, but she soon got used to it.

"There are so many students in each class," she later told Sammy, while they were having dinner in their tiny kitchen. "There's hardly a place to sit."

"What did you expect?" said Sammy. "Nasser made university education free, but he forgot to build new universities to provide space for the multitudes of students who applied."

"I've suddenly become very popular; many students are asking to copy my notes."

"Don't get emotionally involved; you can't help everybody."

"Come quickly," she said, interrupting him. "Put your hand here— the baby is moving."

Sammy put his hands on her belly and laughed as he felt the movements of his child.

"I think it's going to be a boy. He's really kicking," he said.

"I hope he'll look just like you. Wouldn't that be wonderful?"

"It would be wonderful only if I could make more money."

"Don't be discontented. You'd be shocked to see how many of my fellow students are living. Most of them can't even afford to buy books. We're really very lucky!"

"Yes my love, we're from the lucky few, but we still can't afford to be independent. I hate working for pennies."

"I feel blessed to have you, to have a decent education, and more money than ninety percent of my colleagues. Poor Melanie, Hassan won't let her dance or finish her schooling."

"What do you expect from a devout Moslem revolutionary? I feel sorry for her."

"Hassan is a good man and he loves her very much."

"He is too rigid. I don't think their marriage will survive."

"I know you hate him, but I hope you are wrong."

University for Nadia was a great adventure. There was an obvious segregation in the classroom. The girls sat in the front rows and the young men sat in the back. Men and women did not mix, except rarely, and then usually in secret outside the campus. Most of the students had come from public schools and had studied mainly in Arabic. They were from working class families and would otherwise have been forced into some kind of manual labor. The rich students went to the Cairo University in Giza or to the American University in Cairo. Ein Shams was the only option open to Nadia because she was late in registering and because it was conveniently close to her home.

One hot day a young man approached her. He was pale and thin, his clothes were ill fitting, his face was scarred by acne, and his hair was dusted with dandruff.

"Are you Nadia Soliman?" he asked.

She stopped in the shade of a huge acacia tree because the sun was blistering. "Yes, I am," Nadia blushed. Why is he so angry? The young man came so close to her that she could smell the odor of garlic on his breath.

"Do you know that your father's big store is owned by my family?" he asked without any preamble.

"I thought that the store had been in our family for over two hundred years," she answered, blushing so deeply that tears came to her eyes. He made her uncomfortable.

"You're only leasing it from my family." He was fixing her with his fathomless black eyes. Like Coleridge's Ancient Mariner, he was determined to tell her his story. "Do you know how much you're paying every month?" His voice was loud and trembling.

Nadia was afraid of the angry young man. "No, I've no idea," she murmured, thinking of the dangers her father had cautioned her against. She shivered slightly, for he was staring at her with much hatred.

"You pay the whole of four Egyptian pounds every month, less than it costs to buy a few gallons of gasoline for your car."

"I'm sorry, I had no idea." Nadia drew back slightly. She felt physically threatened, but at the same time she also felt sorry for the young man.

"And as if that is not enough, now you are also going to take a job away from a person who needs it more than you do? There is no justice in this world."

"Why don't you raise the rent?" she surprised herself by asking.

He laughed, shaking so much that the dandruff flew off his shoulders. "The rent is fixed because it is in a perpetual trust."

"I had no idea. What does that mean? What can I do about it?" She was suddenly ashamed of her privileged position.

"You can tell your father that a family is starving to help make him rich."

"I'm really sorry. I've really nothing to do with all of this, but I'll tell my father." She tried to comfort the angry young man with a smile, but he gave her a blistering look from his fiery black eyes, turned his back to her, and walked away.

Nadia felt guilty for this man's suffering and frustrated by her ignorance about the world around her. This bitter young man scared her. Later, she told Sammy about her encounter with the suffering Mariner. "What will become of him?" she asked.

"Well, if he gets his degree he will join the hordes of government employees who are barely making a living."

"The revolution is supposed to redistribute the wealth among the people. I'm sure it'll raise their standard of living," said Nadia.

"You're so naïve. Wait and see; the poor will only get poorer and the rich, richer."

"I guess I am naïve."

"Forget about him; most Egyptians are paupers anyway. How about going to the movies tonight?"

"I have to study."

"Damn it. Please, let's go. There's a great Italian movie."

"There was so much sadness in the young man's face," she said.

"I'm not surprised. What about it? The movie, I mean."

"What about me? What am I going to do with my degree?"

"For one thing, you'll be able to recite poetry to your children and read them The Canterbury Tales." Sammy laughed and hugged her gently, trying to avoid squeezing her big stomach.

"You really don't care, do you? You think I'm just playing a game."

"Come on, let's not argue. We'll be late for the movie."

"I don't want to go. I told you I have homework."

He looked at her in dismay. What's the use, she thought. He'll drag me to the movies anyway.

"You can study tomorrow when I'm at work."

"I also want to work someday; I want to be respected for my own accomplishments."

"Wow, wow, what's happening to my little girl? After a few months in college, she's already defending women's rights. Pretty soon you'll be out on the streets rioting like my crazy Aunt Billy and asking for the female vote."

"What is wrong with Aunt Billy? I admire her very much," said Nadia.

"Let me tell you a story. Once, when I desperately needed money, I asked her for help," said Sammy. "She sent me a heavy envelope with her man servant. I got excited and opened it quickly. A bunch of pennies and dimes fell out. I gave all the change to the man, and he left. When I looked into the envelope there was nothing else in it. That was all she was giving me, the miserly cow."

Nadia laughed so hard, tears came to her eyes.

"That's the Aunt Billy you admire so much. Come on, we're going to be late. We'll miss the beginning of the film."

"My God, you're worse than my father," said Nadia.

"How can you compare me to your father? You're so lucky I'm allowing you to do whatever you want," answered Sammy laughing.

"At least my father frankly disapproves, you pretend to be progressive, but the truth is, you're really indifferent. You think I'm a child playing a game.

"Nadia, where is your sense of humor? I was only teasing you, and besides, life is only a game. You're my queen and I'm proud of you. Come on, we're going to be late."

"What's the use?" she muttered. "Let's go to the movies."

A NEW CAREER

A FEW WEEKS LATER WHEN Nadia came home from school, Sammy stopped her in the living room. "Wait here, I have a surprise for you," he said. He went into the bedroom and called her in after a few minutes. The bed was covered with ten-pound notes. She looked up at him in amazement.

"I sold the script. Isn't it just great?" he said.

"It's a lot of money!" exclaimed Nadia, shaking her head, "I can't believe it! You should be very proud," she hugged him. "I always knew you were a genius."

He threw her on the bed and they made wild love upon the crackling banknotes.

"Well, how did you like my surprise?" he said later on, touching her swollen belly lightly with one hand.

"Oh, it's wonderful, I'm so happy for you." In truth, Sammy's new career really perplexed her.

"You don't look very happy."

"No, I am happy, I'm just very tired. The baby is getting to be very heavy."

"It will be over soon," Sammy said. Only two months to go."

"Right after the final exams," said Nadia. As the weeks passed, Nadia continued to study and Sammy immediately began to write another film script. He was rarely home before one in the morning. Nadia was lonely and sometimes scared to be left all alone in the apartment. Nadia woke up one night by the sound of the front door closing. The twirling clock on the dresser, one of six identical wedding gifts, told her that it was three in the morning.

Sammy came in looking pale and quickly ran to the bathroom. She could hear him throwing up.

"Are you sick, my love?" She followed him into the bathroom.

"Don't worry darling, just go to sleep."

"What's this on your shirt? It looks like blood."

"It's nothing. Don't worry."

"Sammy, it's three in the morning. I was . . . I'm worried."

"I had a few drinks at the Fishawi coffee shop with George Aziz, the director. He offered me a role in his new movie."

"A role? You mean acting?" She pointed to his shirt. "It's lipstick! There's lipstick on your shirt."

"Oh! Yes, I guess," he said laughing nervously. "I guess it's lipstick. There was a fight at the bar and one of the guys threw a woman at my chest." He took off his shirt and threw it in the tub. Surely he could come up with a better story; after all, he was a writer.

"Don't you think it's wonderful news?" he asked.

"What, the flying prostitute?"

He laughed out loud. "No, the role, silly! It's too late at night for funny jokes. The role, I mean the role."

Nadia didn't quite understand. Was he having an affair? Was he screwing around? "I'm too tired. I don't understand you. Let's talk in the morning," she said.

When she woke up in the morning he was gone. Who is that man that she married? Life with him was going to be worse than a ride on a galloping camel.

Some days later, Sammy took Nadia to visit Studio Misr. It was an impressive compound of several one-story buildings located among the green fields about two miles from the Pyramids of Giza. Nadia was wide-eyed with amazement; it was a world she had never seen before.

They entered one of the tall, whitewashed buildings and were almost blinded by the bright spotlights. There were two scenes being shot at the same time. The two sets were built back to back and Nadia found it to be strange, walking from a sumptuous living room in a rich man's house to a garret where a student lived.

"The sets look flimsy and fake; I would never have imagined that they were only three sided rooms with thin shaky walls."

"That's the world of make believe, sweetheart." They watched the shooting for a long while. For Sammy, the work was fascinating, but Nadia was bored to tears.

"Can we please go out for a breath of fresh air? I feel dizzy and sick." Was this the place where Sammy spent his evenings when he was supposed to be at the clinic?

A gorgeous woman walked passed them and saluted Sammy. She was Magda Nagy, the famous siren of the silver screen. He returned the greeting with a dazzling but embarrassed smile. He was, without a doubt, a familiar face to her. The actress glanced briefly at Nadia and walked away.

"Sure, let's go out," said Sammy, quickly recovering his composure.

When they reached the exterior of the building, Nadia felt the night breeze filling her lungs with fresh air. In the garden of a distant villa a dog was barking, and she could hear faint voices singing far away. The village of Gorna was nearby, at the foot of the Pyramids. The residents, who in ancient times used to be grave robbers, were now tourist guides and raised horses and camels.

"Please take me home," Nadia said. "I'm tired."

"The next morning Nadia called Melanie. "He's cheating on me," she said.

"Are you sure?"

"I'm almost sure. He says he loves me but I think he's screwing around."

"What are you going to do about it?"

"I think I still love him, I'm going to have his baby."

"Then you'll have to look the other way, and forget his peccadilloes."

"I can go back to live with my parents."

"That would be a lot of fun."

"I feel trapped. What would you do?"

"I don't know. Do you think I have all the answers?"

"Melanie, you seem so wise and always know what to do."

"Let me tell you something funny my deluded pal. Your wise friend has given up her job and her apartment and is going to live with her mother-in-law. What do you say to that?"

"Oh Melanie, what happened?"

"Love, my sweet. Idiotic love. It steals your reason. We are weak females; still depending on the masters of our fate."

"I hate the thought of seeing you kissing other women on or off the screen," Nadia said to Sammy while they were driving in their new white Fiat on their way to lunch at the Hakims'.

"You're such a child," Sammy laughed. "It's only pretending, just acting. Are you jealous?"

"Yes, I'm jealous," said Nadia. "Besides, actors may be well paid, but they're not respected."

"Acting is an honored career. In England, great actors are knighted by the queen."

"Well, we don't live in England. Actors are not respected in the Middle East. My mother says that actors are sinners who'll fry in hell."

She quickly regretted quoting her mother and blushed deeply.

He looked at her and smiled. Nadia's heart flipped; he certainly was handsome and fun to be with, but maybe these were not the best qualifications for a husband.

"Your mother's an idiot," he said. "Do you really believe that actors will fry in hell?"

She wasn't sure what she believed in. She had grown up with the Christian faith at school and Moslem beliefs at home. They had a great deal in common, and yet both sides claimed that they were the only ones favored by God. All this bigotry had weakened her faith in religion.

"No, I don't really believe that you'll burn in hell for acting," she said. "But you'll be tortured by family and friends who think of it as a disgrace, and isn't that hell?"

"I don't care what people say. I just want us to be rich and happy. I'm sorry Nadia, I have to follow my heart."

She had worshipped Sammy but he wasn't her God anymore. "I thought that we were happy," she said softly.

"Of course we're happy! But I want to make a lot of money and I want to do it by amusing people and entertaining them, not by having them pay me only when they're sick. Medicine should be socialized and available for free to all."

"You can't be a socialist and a capitalist in the same breath." She was silent for a moment. "Are you really sure that this acting business is a good idea? We're going to be ostracized by society."

"The hell with society. Do you think any of your rich relatives care enough to pay our debts or buy us this new car?"

"Please don't talk about this to Daddy," Nadia told him. "He's going to be furious."

"No, I won't talk about it, but he'll find out sooner or later."

The scandal broke out a few weeks later when the film posters were put up in the center of town with Sammy's face and name plastered on the walls. A violent storm erupted. Omar Soliman was in such a rage over the news, he almost had a stroke. He went to see the Hakims to discuss the problem.

"When you asked me to give my daughter in marriage to your son, you told me that he was going to be a doctor, and I agreed to the marriage on that basis."

"We all thought that Sammy was going to be a doctor," said Youssef El Hakim. He was visibly uncomfortable.

"So what if he becomes an actor?" said Laila. "Hussein Bey Wahby, the great actor, comes from an aristocratic family."

"Wahby is a pimp," said Omar angrily. "They're all a gang of pimps and prostitutes."

"Please let us discuss this very calmly," said Dr. Youssef. "Sammy is an adult. He has chosen a different path and I believe that we should respect his wish and hope that he has made the right choice."

"What about the army? They'll kick him out in dishonor!" The vein on Omar Soliman's forehead was throbbing.

"He's been given permission from the authorities to do the film," El Hakim said.

"I've never heard of such a thing," Omar's voice was getting louder.

"Has anyone ever heard of a doctor in the army who is also acting on the side? Pimps and prostitutes," Omar repeated.

"I don't think that either of us will ever change his mind," said Doctor Hakim.

The meeting was a disaster and nothing was resolved.

Amina Soliman, after a few days of silence, came secretly to visit her daughter. Her hair had just been done at the hairdresser's and her make-up was fresh. She was a pretty lady and looked elegant in her pale blue linen suit.

"*Ma petite*, we're the talk of the town. I'm ashamed to show my face." Her perfume was so strong that Nadia started to sneeze. "*C'est un scandal!* How our enemies will rejoice!" She continued her litany, wailing all the while, on the theme of betrayal and dishonor. Nadia had mixed feelings; she was both angry at her family's reaction and felt guilty for causing them pain.

"You know, we married you to a doctor—not an actor. He really misrepresented himself. It's a good reason to get a divorce *et sauver l'honneur de la famille.*"

"Mama, what are you talking about? I love Sammy and I'm having his baby."

"It's your choice," threatened Amina. "You'll have to live with this disgrace for the rest of your life! I'm so confused," she sighed. "May Allah forgive me. I don't know what to do. Your father is blaming me for everything." Amina started crying bitterly.

"Mama, it's not your fault. Please, be brave; this crisis will pass. You'll see. Father will calm down. He's a reasonable man." As was her habit, Nadia assumed the role of advisor when she was with her mother.

"Your father doesn't want me to visit you anymore."

"What are you saying?"

"He wants to teach Sammy some sense by breaking all relations," said Amina through her tears.

"What about me? Are you telling me that he wants to disown his own daughter?" said Nadia, her voice rising in anger.

"I don't know." Amina continued whimpering. "He's so angry at the moment. Maybe he'll relent, but right now he doesn't want to see you or Sammy."

"I can't believe that he would do that."

"I have to go. He doesn't know that I'm here. Here, baby, take this."

She gave Nadia a hundred pounds that she must have pinched from the household expense account.

"Mother, I don't need the money."

"Just keep it. May God heal his anger."

"I love you Mommy, don't desert me."

"I love you too. I have to leave; I'll keep in touch." Amina ran out before she could break down and cry again.

The phone rang and Nadia picked it up. There was a woman at the other end. "Can I speak with Sammy?" she said.

"He's at the studio," answered Nadia.

"No, he's not. Tell him Magda called and that I miss him."

Nadia hung up. She felt sick and ran to the bathroom to throw up.

When she felt a little better she called Melanie.

"What am I supposed to do?" she asked her.

"Nadia, just confront him and ask him to explain himself. There must be an explanation."

When Sammy came home Nadia confronted him. "Sammy, Magda, the movie star called for you. She says she misses you."

"The bitch. Just ignore her Nadia. She's nothing but a whore."

"What do you mean? Are you saying she's inventing things?"

"Sweetheart, you are such a child. I love you with all my heart. I work with this woman and she is a troublemaker. What do you expect?

She is just teasing you. You have to trust me. I will always come home to you. No matter what happens outside our home. Do you understand?"

"I'm afraid I do."

THE SUMMER OF '55

WHILE SHE WAS DRESSING up, getting ready to go out, Melanie compared her life with that of her friend Nadia. It was strange to see Nadia, who had been conditioned to accept and obey, now beginning to think of rebellion, while she, the free-thinking Melanie, was sinking deeper every day into self-renunciation. There was always a dark presence lurking in the corners of her existence, like a beast about to pounce, and she couldn't get rid of it. When she was with Hassan, she forgot all about her fears. But when he left early in the morning to go to work and she woke up to find herself alone, she could feel the evil threat. Her uncertainties made her surrender her will to Hassan.

When she had confessed to him her fears that she might be losing her mind, he had just laughed at her and told her that she should stop reading all those psychology books that were driving her crazy. He'd never take her seriously, just like that psychiatrist she once visited, who prescribed sedatives and tried to kiss her in his office.

Melanie took off her red sleeveless blouse, her favorite, and put on a long-sleeved beige one. It made her look like a high school student, and it would surely make her sweat, but Hassan would be happy that her sinful arms were not showing. She also put on a flowered skirt that came down to her calves and tied a beige leather belt around her waist.

That should be demure enough, she thought. What difference did it make? She only wanted to make him happy.

She couldn't believe that it was already two months since they were married on a scorching July afternoon. It was a private religious ceremony at her father's house. The old bastard gave her away as if she were his property. Many of Hassan's relatives attended, but on her side, there were only a couple of uncles who were almost complete strangers to her. There were also two doctors, friends of her father, with their British wives in tow. And then there was Nadia and Sammy.

She was glad that the folklore group was touring in Europe; it would have been hard to explain why she hadn't invited them. As for her mother, she was busy with her second husband somewhere in South Africa. She hadn't received a letter from her in six months.

Melanie was shocked to find out that Hassan paid five hundred pounds for her and also promised to settle with her, in case of divorce, another sum of a thousand pounds. "You're a bargain!" Hassan had teased her. She told him that she would have preferred to get five cows and six goats.

Hassan was still in the shower. She sprayed some *Arpège* perfume on her neck and wrists and wondered if it was another sinful thing that she was doing. According to Islam, women couldn't be trusted; their bodies had to be hairless, sexless, and scentless, or they'd wreak havoc on the fragile male psyche! She was ready and waiting for Hassan to finish his prayers and dress up.

She went out of the bedroom to the living room. She was relieved not to find her mother-in-law sitting there. Actually, the woman was very discreet and left them alone. Melanie picked up the book on ancient Egyptian magic that she had found in an old bookshop in Luxor and started to read. Maybe she could cast a spell on Hassan and make him stop praying so much.

"I don't like that guy," he said, as he came out of the bedroom, ready to go. "Do we have to see them?"

Tonight was the first time they would go out with Nadia and Sammy since they'd been back from their week-long honeymoon in Upper Egypt.

"Hassan, she's my best friend!"

"Yes, my love, I know. Just thinking aloud. I hope the film is good."

"You know what? He doesn't like you either," she laughed and put the book away. Does he know how much she loved him? She reached up and caressed his fresh-shaved face. He smiled and carried her in his arms towards the door of the apartment.

"I love you," she told him.

Hassan and Melanie parked their car in front of their friends' building. Hassan honked once to signal their arrival. Melanie saw Nadia wave from the balcony.

"Good, they're coming down," she said. They were like high school kids going out on a double date. Poor Nadia had never been on a double date in her life. Melanie again felt a pang of guilt for having helped her sneak out of school to meet Sammy secretly. At least she loves him; better than an arranged marriage, or is it? How arrogant of me to have tried to influence her.

Nadia came out of the building wearing a sleeveless pink maternity dress and carrying a white shawl on her arm. She looked like a ripe fruit ready to burst.

"You're blooming," said Melanie, as she kissed her on both cheeks.

"Yes, she's ballooning," said Sammy laughing. Melanie gave him a dirty look. "Just kidding," he laughed.

The men were both wearing gray pants and white shirts. Hassan was all buttoned up and carried a gray pullover in his left hand, but Sammy had his shirt unbuttoned almost down to his waist and had a bright red alpaca cardigan thrown across his shoulders. Melanie smiled, remembering how she had once compared him to a peacock because of his colorful get-ups.

The four of them walked together towards Cinema Normandy, one of the two open-air cinemas in Heliopolis. The women, as usual, fell a few steps behind the men.

"I'm sorry; I'm forcing you all to walk," said Nadia.

"Don't be sorry; we all need the exercise," said Melanie.

Melanie was familiar with the street. It reminded her of her mother when she used to accompany her on visits to friends who lived in the area. Mother, mother, why have you abandoned me?

"How's your Mum?" she asked Nadia.

"She sneaks out to see me once in a while. She's rather pathetic."

They're all pathetic, thought Nadia.

"That's better than my mother who doesn't even bother to write."

"Well, it's really not too bad. After being suffocated by too much attention, I'm suddenly alone. I discovered a kind of freedom, a space for myself, but to be honest with you, I miss my parents very much."

Nadia noticed that Melanie was beginning to sweat in her long sleeves.

"Why are you all covered up? It's too hot."

"I thought it was going to be cool in the evening." Melanie blushed.

Nadia was silent. They passed in front of Christo's Toy Store; its half empty shelves seemed to be haunted by ghost toys. Christo, a skinny ghost himself, sat in front of his store. There was a sense of decay about the place. Ali's Barbershop was still busy as usual, full of clients sitting around and chatting. But next to it, Aram's Photo London House was shuttered. It had once been popular with the British military.

"Almost every person in Heliopolis had his picture displayed in that window at one time or another," said Nadia remembering how her own picture had been sitting there collecting dust for six months.

"Yes, it was like a family album."

As they passed the food stall close to the cinema, the smell of frying *falafel* and roasting *shawerma* assailed them. People stood in line to pick up their sandwiches. Melanie had hardly eaten anything all day but she felt no hunger. When they caught up with the men, Nadia asked, "What are we going to see?"

"I've no idea," said Melanie.

Nadia looked wonderful in spite of her swollen belly, thought Melanie. I'm saddened at the thought of being childless in spite of the fact that I don't want any children. "I'll never have a child." she thought.

"It's called *Back Window*, no, no, it's *Rear Window*, I think, yes," said Sammy. "Hitchcock . . . should be good. Grace Kelly and what's his name . . ."

There wasn't much choice available, but seeing movies was still one of Sammy's favorite summer rituals. At least Hassan and Sammy won't get a chance to argue about politics, thought Nadia.

"I love Grace Kelly," Melanie said.

"You know that you resemble her? The same fragile air," said Nadia smiling.

"Really, you think so?" Melanie didn't think she resembled Grace at all. She didn't believe that she was fragile either.

"Oh definitely, don't you agree, Hassan?"

Hassan examined Melanie's face for a moment.

"Yes, they do look alike; they're both so very skinny."

"Thanks a lot," said Melanie laughing.

"Jimmy Stuart, I just remembered his name," Sammy said.

Melanie told herself to eat more just to please Hassan. He thinks she's too thin.

"Now, tell me more about the honeymoon," asked Nadia.

"Great, but one week was too short."

"Our honeymoon will last forever," Hassan said.

"He'll not put his arms around my shoulders in public, but he loves me fiercely in private. He prays to Allah five times a day and would be shocked to know that he was my God, my prophet and my religion."

Melanie smiled at him.

"Just beware of getting pregnant," Sammy told Melanie. "It'll sure put a fly in the honey pot," he laughed aloud.

Nadia blushed. Sammy not only mangled his clichés but also was sometimes incredibly coarse. They all laughed anyway.

They passed in front of a strange store whose doors were always wide open. It sold slabs of compressed animal manure, which were used to burn in the ovens instead of wood. The smell was overpowering. It never failed to make Melanie smile seeing the "shit" store.

When they finally arrived at the cinema, people stared at them; they whispered and smiled. A young boy came up to Melanie and asked for her autograph. Hassan attempted to send him away, but she stopped him and signed the flyer the boy extended to her. Many of the women were eyeing Sammy and Melanie with interest.

"I keep forgetting that you and Sammy are famous stars." Said Nadia.

"We'll just pretend that we don't know these two, shall we?" Hassan answered.

He was trying to act cool, but Melanie knew that it irked him that she attracted attention. She touched his arm lightly trying to reassure him that she was completely his. She couldn't believe how much she had given up for this man. She left the folklore group for his sake. She was wearing long sleeves and a long skirt for his sake. All this she did for the security of being in his arms. What more could she do? He should know that the public had a very short memory. They'd soon forget her.

"We'll just have to get used to the limelight," said Sammy, running his hand through his curly hair and flashing his perfect teeth. He thrived on the attention.

When the men went to stand in line to buy the tickets, the women remained on the side waiting.

"I miss the theater very much," Melanie said softly to Nadia; "People will soon forget Soraya Salem and I'll be known only as Mrs. Hassan Hosny, a very loving and devoted wife."

"Do you regret that you renounced the stage?"

Melanie hesitates, "I'm happy now, my life depends on Hassan."

Nadia was shocked at the change in her friend. "I confess that I didn't expect you to give up your independence so easily. Are you not scared?"

"Yes, for more reasons than one."

"What reasons?"

"Too complicated."

Nadia was hurt that Melanie was so secretive. "I miss the foulmouthed girl that I knew so well. Damn it."

"Relax. I'm playing the good wife. You think I'm a hypocrite?"

"Come on, Mel, you're just in love."

"Give me some time. I'll go back to swearing very soon. Everything is still very new with us two. I'm still in a trance."

"A trance?" Nadia laughed, "He seems to have knocked you on the head with a mallet."

"Shush, I'm very happy," said Melanie in a low voice; "I've no regrets."

After a moment of silence, in which they stood together, their shoulders touching and their eyes fixed on their spouses, Melanie whispered, "I had my fortune read by an old woman when we were in Aswan."

"Really, what did she say?"

"She foretold an imminent death in my circle."

"I hope you don't believe in that stuff! You must have paid her too little." Nadia laughed. "Remember when you showed me that smashed mirror in school?"

"How can I forget?"

"You were so terrified of having seven years of bad luck. Well, look what happened, a few years later you're married to your true love."

"The seven years aren't over yet Nadia."

"Stop this craziness! I can't believe that you're so superstitious!"

The men finally returned with the tickets and they all entered the cinema together. They found themselves in the familiar open-air space with the plain large screen facing them. The two sidewalls were covered with greenery that softened the bare look of the place. It was six-thirty in the evening and still not dark, but the rows of cane seats were quickly filling up with people.

There were children running up and down the isles, babies crying, young ladies with their chaperones, couples, friends and neighbors.

It seemed as though the whole community had decided to go to the movies. Vendors of *Semeet* were walking around selling sesame bread pretzels, boiled eggs, sodas, and lemon ices.

The two couples settled themselves in a center row. The weather was very warm, for it was mid August and not a single cloud could be seen in the light blue sky. There was a faded crescent moon hardly visible just above the screen.

"*Hel helalak, shahr embarak!* A new moon, the blessed month of *Ragab* is starting!" said Hassan and quickly kissed Melanie on the cheek for good luck. "It's a sacred month because if you fast a single day in this holy month, one of the many doors of hell will be closed to you."

"I didn't know that," said Melanie. She looked at Nadia, thinking that here was an example of real superstition. How many doors does hell have? She prayed that Sammy didn't decide to make a joke about heaven and hell. Luckily he didn't, but he turned to Hassan and asked him, "What did you think of the president's speech?"

"I thought it was eloquent but not very diplomatic," said Hassan softly, after some hesitation and looking around him nervously.

"Diplomatic! 'There will be no peace on Israel's border because we demand vengeance, and vengeance is Israel's death.' How diplomatic can you get?" said Sammy in a loud voice.

"He gets carried away with the rhetoric. He feels very deeply about the rights of the Palestinians," said Hassan calmly.

"But this, this is irresponsible," insisted Sammy.

"Where exactly are Israel's borders anyway?" asked Nadia.

Hassan explained, "Well they have Jordan and the west bank to the east, Syria and the Golan heights to the north east, and Lebanon to the north. Egypt and Gaza are on their south-western borders."

"That means they are surrounded by Arab states. It's a suicidal choice," said Sammy.

"The promised land just happens to be in a bad neighborhood," joked Melanie.

"I hope there won't be another war," commented Nadia. The baby gave a great kick as if he also was protesting.

"The Security Council is busy trying to find a settlement but emotions are high on both sides. We might be heading towards another war," said Hassan. Nadia thought that he sounded just like the official news broadcast.

"At least for tonight, Hitchcock will keep our minds off the war for a couple of hours," said Sammy. "Hey, would you all like something to eat before the movie starts?" He was already on his feet. Almost everyone around them was eating.

"How about a falafel sandwich and a lemonade?" said Nadia, who was always hungry.

"The same for me, please," Melanie decided that she had better eat.

"Let me go and get it," said Hassan, "What would you like, Sammy."

"The same, I guess . . . I appreciate it. I would go with you but I don't want to miss a single scene from the movie."

"Are we going to see the Arabic version soon?" said Hassan, taking a jab at Sammy as he hurried away.

"Why not?" Sammy told Melanie, "If it translates into an Egyptian story, we'll see it as an Arabic film before the year is over." Sammy didn't disapprove of plagiarism.

"Hassan is being sent to Russia," she said.

"What! What are you saying?" said Sammy.

"We're leaving in a few weeks. He's been appointed assistant military attaché to Moscow."

"That is a surprise," said Nadia.

"We weren't sure till yesterday. It's confirmed and we leave in about two weeks."

"Congratulations, it must be exciting!" said Nadia, but was shocked and upset that Melanie had not confided in her. Were they really drifting apart?

When Hassan returned, he distributed the pita pocket sandwiches and the drinks. The smell of the *tehina* sauce was strong and pungent.

Hassan passed around the pickled radishes and onions in separate paper cones.

"The film hasn't started yet?"

"Most probably," said Sammy, "the copy is being held up in traffic on its way here from downtown Cairo. By the way, congratulations on your new position," Sammy said.

"Thank you, it was very unexpected." Hassan looked at Melanie.

"I read that the American State Department has confirmed the arrival of several shipments of arms from the Soviet block to Egypt," said Sammy.

"Maybe, I wouldn't know," replied Hassan scowling.

"The deal is supposed to be an exchange of cotton for arms. I guess we're falling under the influence of the Russians."

"The Americans refused to give us a loan for the Aswan Dam; are you surprised that we have to look towards the east?" said Hassan.

"Anyway, remember that Egypt is a Moslem country; we're never going to become communists."

The film finally started and that put an end to the discussion.

Melanie took a deep breath and a big bite of her sandwich.

On their way back, they walked slowly. The temperature had fallen about twenty degrees, as it always did at night in the desert. Melanie shivered and put on her jacket.

"I have to see this movie again to study the technique of Hitchcock," said Sammy.

"Grace Kelly is so beautiful," said Nadia.

"Did you see how the camera moved? One day, I'll direct my own movies," declared Sammy.

"I enjoyed it, got carried away with the story," said Hassan.

"You won't be around when the baby is born," Nadia said to Melanie. "I'll be all alone."

"You won't be alone. Sammy will be with you and I'm sure your mother will be beside you too. You'll send us pictures and we'll see the baby when we come back on holidays. We'll be back in a few months."

Suddenly, they were all quiet, each one engrossed in his own thoughts. The women unintentionally lagged behind the men again, in the good old Middle-Eastern fashion.

A few days later, Melanie went shopping with Nadia for a heavy coat and gloves that would be necessary for the Russian winter. They avoided discussing politics, but they both knew that the tensions in the Middle East were very high and that any day now a war might explode.

On every corner they could see armed soldiers standing in readiness. The entrances to apartment buildings were barricaded with sand bags.

The streets were full of armored cars carrying soldiers to and from the barracks. They went to *L'Americain Cafeteria* for a sandwich and an iced coffee.

"Your hands are icy and you look pale. Are you pregnant?" asked Nadia.

"I'm fine; I'm slightly anemic but nothing serious. And definitely not pregnant," said Melanie.

"You can't fool me. Something is wrong."

"I'm in love, I'm as happy as I'll ever be. Don't ask me what's wrong; things are never perfect."

After a moment she said, "He's a devout Moslem and is desperately trying to convert his Atheist wife. Nothing important, nothing to worry about." They both laughed uneasily.

"How does the old lady treat you?" said Nadia.

"She welcomed me with warmth but with some diffidence. It startled her that I spoke Arabic and knew how to cook *Kishk* and *Mulukhia*. She had expected a foreigner and was pleasantly surprised that I wasn't a *khawagaya*. We get along fine."

"I'm glad; I was worried that living with your mother-in-law might be a problem."

"We have very little privacy but we're leaving in a few days anyway.

The only thing that irks Mama Wafeya is that I don't join her in her prayers. I overhear her constantly praying that God may show me the right path. Actually, it's a bloody good thing we're leaving." Melanie laughed.

"It must drive you crazy," said Nadia.

"Actually, I'm already bloody crazy," shrugged Melanie.

"Let's make plans for the four of us to go out for a farewell dinner."

"Okay. I would like to get drunk and dance all night. Maybe you can slip me a vodka tonic under the table," said Melanie.

"I'm going to miss you."

"No you won't. You'll be so busy with the baby and your classes; you won't have time to think of me freezing my ass in Red Square."

A few days later the four of them met at the Mena House Hotel for a farewell dinner.

"Mena House Hotel was built in 1869 in Giza, at the foot of the Pyramids." Sammy read aloud from the front of the menu. *"The luxurious hotel has hosted Churchill, Roosevelt, Chiang Kai Chek, the Agha Khan, and Agatha Christie, among others.* Wow, it's the perfect setting for a film before the Second World War."

"By the way, I was assured that they have central heating in the apartment where we're going to live," said Melanie. "So we are not going to freeze to death, after all." She was leaning against Hassan and held his arm protectively.

"Imagine, Melanie, you'll be able to see the Russian ballet and listen to the Moscow Philharmonic as much as you want!" exclaimed Nadia with enthusiasm. "It will be very exciting."

Sammy looked at his wife tenderly. She was always eager to make other people happy. He squeezed her thigh.

"The most exciting thing is that we'll be representing Egypt during a very crucial period of its history," said Hassan.

Mercifully, the food arrived and the conversation paused for a while to be picked up later by Sammy. "What exactly are you going to do in Russia?" Sammy asked. He loved to challenge Hassan. Hassan sighed and remained silent for a moment as if gathering his thoughts. It was a habit of Hassan's that unnerved many people and irritated Sammy.

"I guess I'll do what diplomats do all over the world, attend cocktail parties and try to represent my country in an honorable way."

"Melanie can become a spy like Mata Hari and reveal to us all the dark secrets of the Kremlin," Nadia said.

"Don't give her any ideas," said Hassan, loosening up slightly and almost breaking into a little smile.

"I'm sure that it is a very interesting job," said Nadia. "How long do you think the mission will last?"

"The mission is for three years, but we'll come back at least once a year to visit."

"I'll miss you very much," said Nadia.

"We'll stay in touch and we'll be back before very long," Melanie reassured her.

"I'll send you pictures of the baby."

Sammy started rubbing the side of his nose with his left hand. It was a nervous habit he couldn't break. "Why does the revolution have to demolish everything that came before it?"

"The old system was corrupt and had to go," said Hassan calmly.

"Evolution is far better than dissolution," continued Sammy. He was unwilling to be diverted from his previous train of thought. "I distrust revolutions," he insisted.

"Our revolution is different; it is based on ideals of equality and a better life for the people," said Hassan, with conviction.

"A military regime will always be subversive. You're never going to convince me that fascism is good for the people," replied Sammy.

"We're not a fascist government." Hassan was getting angry. "We'll soon have elections and representation in parliament."

"Hey you guys, stop arguing; let's drink to our future," said Melanie, raising her glass of orange juice.

". . . And to good friends," said Nadia.

"To good friends," repeated Sammy.

". . . And to the United Arab Republic," said Hassan. He was drinking soda water in a wine glass.

Long live the revolutionary fanatics, Sammy thought to himself, as he gulped down the rest of his whisky. I really dislike everything about him, including his sexy voice and his green eyes.

18

A CHILD IS BORN

NADIA SAT ON HER narrow balcony overlooking the graveyard that lay across the street from her building. She had her feet up on a chair and was sipping a large lemonade. The sun was liquefying the black asphalt on the street and making it glow like spilled ink. The ficus trees lining the road were covered with the desert dust, and instead of bright green, were grey.

She watched, without much interest, the street that was melting like molasses. The heat was unbearable; it sapped all her energy. She moved her swollen body around to find a comfortable position, but it was no use. Whatever she did she couldn't be at ease; a tyrannous intruder, who delighted in torturing her, was occupying her body.

She saw from half-closed eyes, a woman who was coming towards her building. She was walking gingerly; her high heels sinking into the soft ground. The molten street forced her to step awkwardly and then, when both her heels got stuck in the melting tar, she toppled over like a broken doll. Nadia watched her as she picked herself up and continued walking across the street.

She recognized Shoushou, the woman who lived on the first floor with her husband, an army officer. Once, when Melanie visited her, she saw the man slap his wife and drag her into their apartment. He screamed and called her a slut. Shoushou was now a widow and was wearing black.

Poor Melanie, I wonder what she's doing in Moscow? Between the chatty lines about the *Corps Diplomatique* and their cocktail parties and high teas, Nadia could detect trouble. Melanie was uncharacteristically vague, avoiding any mention of private issues between Hassan and herself. "I have a feeling that Melanie isn't happy. I'll have to write to her more often," decided Nadia.

The doorbell rang. "Who could it be? No one comes to see me these days; I'm a social outcast," she thought as she shuffled towards the door.

Ne'na'a, who came once a month to wax the hair off her body, was standing at the door breathing heavily. Her huge body was wrapped in a black silk *melaya*, but under the dark sheet Nadia could see a yellow dress bursting with flowers of every shade and color. "*Ya Binti*, when in God's name, my child, are they going to put in the blessed elevator?

This climbing up the stairs is cutting my heart to pieces." Her face was red and her breath was short from the effort of climbing three flights of stairs.

"Sorry, Ne'na'a; please come in and rest yourself."

They both waddled to the bedroom and the woman collapsed on the carpet in a huge pile of flesh and flowers. A yellow scarf decorated with multi-colored beads, from which two jet-black braids emerged, was

tied tightly around her head. Ne'na'a was around fifty years old. Her round face was smooth and shiny. There were no wrinkles to mar her rosy cheeks. She folded her black *melaya* by her side. Nadia ran to the kitchen to get her a glass of cold water.

"They promised that the elevator would be in before the end of the year, but you know how it goes!" said Nadia. She watched Ne'na'a drink the water noisily and slowly regain her composure.

"*Bokra fil Mishmish*," said Ne'na'a laughing.

"I guess you're right; tomorrow when the apricots ripen is an empty promise," Nadia smiled.

The woman sat on the floor crossing her legs in a Buddha stance and took out a package wrapped in brown paper from her sack. She opened it carefully and extracted the hardened sugar and lemon paste that she had prepared early that morning in her home. She broke off a piece and began kneading it in her plump hands until it stretched out like taffy and became white and malleable. Nadia got a large towel from the bathroom and spread it on the carpet. She slowly took off all her clothes and sat nude opposite the woman on the towel. Ne'na'a put her hands on Nadia's belly, firmly but gently.

"It feels as if your time is near."

"Yes, it's anytime now," said Nadia, blushing uncomfortably under the woman's scrutiny. Ne'na'a smiled her good-natured smile. Her lovely black eyes were encircled with black kohl and looked like huge saucers.

"Lie down my child and open your legs. We'll begin from down there. Let's get the difficult part over with first."

Nadia got a pillow from the bed and lay on the floor with her legs open; she felt embarrassed but reassured herself that it was no big deal for Ne'na'a. She was the grooming woman, the *Ballana*, who was allowed into the homes of the best of families. She knew all the intimate secrets of the women and the girls and spread gossip as she spread her sugar paste. She also doubled as a matchmaker and was responsible for many arranged marriages. She was an institution, not a person.

Nadia was apprehensive because the removal of her pubic hair was always excruciatingly painful. She was never going to get used to this ritual. Melanie refused to subject herself to the painful procedure, but it had never occurred to Nadia to rebel. "I just have to relax and endure it," she told herself. "It was a Moslem tradition and it had to be suffered as best as one could. Or was it an ancient Egyptian custom? Whatever!" thought Nadia.

"Please be careful Ne'na'a, last time you took the skin off."

"Don't even mention a thing like that! God preserve us, God preserve us. Don't you know that he who's afraid of the devil is bound to encounter him? May evil stay outside and far away!" She spat down the front of her loose gown in order to ward off the evil eye.

"By the way, do you know Mervat Abu el Naga?" said Ne'na'a. Her spicy perfume was making Nadia slightly nauseous. The woman used a local perfume made of jasmine, musk, and amber; it was overpowering.

"Yes, I know her very well; her husband was— ouch, you're hurting me—he was a friend of my husband in high school."

"Well, they returned from America last week."

"So soon? I thought that Zaky Abu el Naga was going to get a Ph.D. in Physics?"

"It seems that she just hated America. Apparently, the women there work like slaves. They live alone, away from their families, with no one nearby to help them. It's a hard life. She just couldn't survive without being the center of attention and having all her family and servants around her."

"She's so spoiled. I'm sure that America has some exciting things to offer. Women there are free and equal to men."

"Maybe, but Mervat felt she was a nobody in America. I don't blame her. Here, she is important; a hen in cream sauce. She forced him to return."

Nadia laughed, "Poor Zacky; she's ruining his career."

"They're so rich; why does he need to be exiled from Egypt? Who wants to be in America? Egypt is the mother of the world."

"I'd love to go to America. I want to get a degree and become a professor."

"With a baby on the way and others to follow, God willing, you'd be lucky to finish college."

Nadia lay down silently and endured the agony of the depilation, which left her pubis tingling with pain and her eyes full of tears.

"You never know," she said. A silence followed, interrupted only by the sound of the sugar paste being spread and then ripped off Nadia's limbs. Working on her arms and legs was relatively painless and gave Nadia a chance to relax. She remembered that Ne'na'a regularly visited the wife of the American Ambassador. She was announced as Mrs. Peppermint because Ne'na'a, which means peppermint in Arabic, was unpronounceable in English.

"Well, I'm finished," said Ne'na'a, "do you want me to give you a bath?"

"No, thank you. I'll do it myself." Nadia quickly got up, covered her body with a blue cotton robe that lay nearby on the bed, and slipped her feet into soft terrycloth slippers of the same color.

"Suit yourself. Mervat loves for me to scrub her down with the black woolen loofah. Her skin glows like marble."

"Thank you, my dear, maybe next time."

"Whatever you want. May Allah pull you through your ordeal safely. Enshallah, I'll see you next month, God willing."

"*Enshallah*, thank you, Ne'na'a. Let me get your money."

Nadia opened the top drawer of her dresser and pulled out a pound note. She helped the woman off the floor and handed her the money.

The woman slipped the bill into her bra and slowly shuffled towards the bathroom, where she disposed of the sugar paste in the toilet and washed her hands carefully. The whole house now reeked of her perfume.

"I leave you in the hands of Allah my dear."

Ma'a el salama; go with peace," said Nadia.

Nadia closed the door to the apartment softly and walked back to the bathroom, where she took a nice long, hot shower to ease her aching hairless limbs. She felt a little sorry for herself. She was so alone she might as well be in America! She sat down at her desk and wrote Melanie a long letter, but the letter made her feel even lonelier; she needed her friend by her side.

Her mother, who snuck in secretly to see her every once in a while, was not much help. She was so afraid of everything; Nadia found herself comforting her and not vice versa. She missed Melanie more than anything else.

Sometimes she wanted to tell Sammy how she really hated being alone, and disliked having her family avoid her after she was the center of their love. But she never could bring herself to confess to him her secret thoughts. It would only disturb him. It was best not to show any signs of unhappiness, for she knew that Sammy would feel guilty and be on the defensive. He would only scold her for being weak.

She tried not to feel hurt that her relatives did not come to visit her anymore. They were stupid to think that being an actor was a disgrace, but she could not change them. She wanted them to love her in spite of the fact that she was married to an actor and was the subject of gossip. Sammy didn't care what anybody thought. "Are you going to be like your mother, worrying all the time about what people say?"

he told her once when she complained. He swam against the current of public opinion, sure of himself, undeterred by obstacles. She had to admire that.

She, on the other hand, had a hard time facing her fellow students' curious glances. The whole world seemed to either pity her or despise her. What saved her from total disaster was the love for the baby moving inside her and losing herself in her studies.

Nobody explained to her anything about childbirth. Her mother and her mother-in-law were busy preparing the clothes for the baby, but not one word was said about the event itself. The gynecologist measured her pelvis once and declared that she would have no problem. She had a blood test done when her feet started to swell, and was advised to walk an hour every day and drink a lot of water.

She was only nineteen and ill-prepared for the experience. Desperate to find out what was going to happen to her, she borrowed a medical book from her father-in-law's library and read about childbirth. The pictures and the descriptions of all the technical procedures frightened her to death. All the complications were discussed, but there was nothing available from the point of view of the woman. It was as if the doctor would be doing all the work. Sammy caught her reading the book and snatched it out of her hands.

"You have no business reading this book. It's only for doctors. You're young and healthy and this book deals only with problems," he said.

"I would like to know what to expect," Nadia replied.

"Sweetheart, it's a natural process. You'll have cramps and then the baby will be pushed out. Nature is going to do the job for you."

"But Sammy, I've heard horror stories."

"Women like to make a big deal out of it. Don't worry so much. Do you know that the peasant women squat down and have their babies in the field, cut the cord with their teeth, tie it, and continue on with their work?"

"Some women die in childbirth. Emma Bovary and Anna Karenina almost died."

"That was 100 years ago! You'll be okay. I'll be with you. You'll have a good doctor and a hospital staff looking after you. You'll also have anesthesia, so you won't feel the pain. We're in the twentieth century for heaven's sake. Relax."

A few weeks after Nadia had taken her final exams, she went into labor. Sammy took her to the movies as soon as the contractions started.

They watched the movie while he monitored her contractions. The pain seemed bearable. After the movie, they went for a walk until the pain made it impossible for her to continue. He finally took her to the hospital. The doctor was there waiting for them. "Where have you been? I've been worried about you. I've been waiting since you called me three hours ago."

"We were at the movies," said Nadia feeling guilty.

"She wasn't ready," said Sammy calmly. "It was a good film."

Nadia was soon crying and screaming from the pain; the blood was covering the bed sheets, and she was sure that she was going to die. She thought that God was punishing her for her sins. Her mother, who had come to be with her, was of no help. All she could do was cry. Sammy's mother came into the hospital room and whispered to her:

"He who licks the honey shouldn't cry when the bee stings. Nobody forced you or held you down when he was on top of you, so stop screaming now."

Nadia was shocked into silence. She felt as powerless as the sacrificial lamb that the cook butchered every year for the feast of Bayram. Who were these masked men surrounding her? They poked and pulled at her while she lay at their mercy, bleeding to death and screaming.

The numbing gas finally arrived in a huge tank. They gave her a mask to breathe it in. She took a deep breath and waited eagerly for oblivion, but nothing happened.

"There's nothing in the tank," she yelled. But the doctors only laughed. She said "Damn you, help me!"

Sammy took a sniff from the mask and discovered that the gas tank was empty.

"She's right. Get another ether tank, quickly."

"We need to do an incision," a voice said.

"No, I don't want an incision," Nadia screamed. The pain was so unbearable she fainted.

By the time a full ether tank arrived, the baby had already been delivered. It was a boy!

The nightmare was over. As soon as Nadia held the baby in her arms, she forgot about the pain. He looked just like Sammy and she loved him completely. His little hands held on to her fingers and his eager mouth sucked the milk from her breast. He was so tiny and he smelled so good.

"What should we call him?" asked Laila.

"I want to call him Ahab," said Nadia.

She was still angry, not only with her mother-in-law but also at the doctors for treating her so callously. She couldn't forgive her mother for not telling her the truth about what to expect.

"What kind of a name is that? I think we should call him either Youssef or Omar like one of his grandfathers," insisted Laila.

The argument continued relentlessly for seven days. Sammy told her she could call him whatever she liked. Nadia insisted on Ahab and finally she got what she wanted.

A grandson was too much to resist; the Solimans relented and reconciled with their daughter, but their relationship with Sammy remained very strained. Every member of Nadia's relatives continued to think of Sammy as a reckless and irresponsible man.

She couldn't get enough sleep. The baby cried constantly. She had to feed it at all hours. Sammy moved to the nursery and let her sleep with the baby. Her old nurse, Dada Baheya, came back from the village to help her out. Sammy was impatient when dinner wasn't ready, and he was furious when she met him in her milk-stained nightgown. He started eating out.

The two grandmothers couldn't agree on anything. They confused Nadia with contradicting advice on almost everything that concerned the baby. The only thing that they agreed upon was that Dr. Spock didn't know what he was talking about. Nadia gave in to Laila, who swaddled the baby in a white cotton sheet like a mummy and insisted on rocking him to sleep. She agreed with her mother that she should feed the baby whenever he cried, and not on a schedule like the American doctors advised. It seemed to Nadia that she had been turned into a milk machine, and when Laila said one day, "Thank Allah our little cow has enough milk," Nadia broke down and cried. Sammy didn't help; he always came home late, looking tired.

"I hardly see you. You're working too hard," said Nadia.

"I have to juggle all three jobs; my job in the army, the clinic at El Merg and my acting career. Pretty soon I'll have to make a choice. Just be patient."

Sammy didn't heed his own advice. Too impatient to wait for her incisions to heal, he forced himself on her while she was still recuperating and ripped her vagina open. She was rushed to the hospital to be stitched up again.

Nadia's sense of vulnerability turned into a post-partum depression that lasted many weeks. She slowly healed and started feeling better.

The baby grew quickly and a new term at the university started. Dada Baheya promised to stay and take care of the child. The time passed quickly between the baby and her university. Sammy was hardly ever home.

A year after their son was born, Sammy came home one evening with a gift behind his back. He gave her the small package. It was a ring with a three-carat diamond.

"Have you robbed a bank?" She had never seen such a large diamond in her life.

"I sold my Balzac script. You're looking at the writer and producer of *Cousin Fatma.*"

"How wonderful," she laughed. It seemed that money was going to make them happy after all.

MOSCOW

MELANIE WALKED OUT OF the Egyptian Embassy on 13 Kropotkinsky Street along the outskirts of Moscow, and into the cold avenue. She was wearing two pullovers and a heavy woolen coat, but the chilly wind still penetrated her bones. The roads were slippery with ice and veiled in a milky vapor.

"*Prasteet-ye*, Sorry!" she said to the pale and angry woman who absentmindedly bumped into her.

"*Astarozhna*, Look out!" the woman yelled back.

"*Ya ochen' sazhilyeyoo*, I'm really sorry!" said Melanie trying to practice her Russian, not sure if she was pronouncing the phrase correctly.

"*Ookhadeet-ye*! Go away!" the woman answered angrily; her threadbare coat was pathetically inadequate for the sub-zero weather.

Melanie forgave her instantly because she felt sorry for the Russian people whose lives were unbearably difficult. It was almost Christmas but there was no cheer in the Russian gloom. Frozen people with pale faces and heavy bodies were dragging themselves around to unknown destinations. A Russian never smiled until after six in the evening when he was filled with Vodka and had reached a point of alcoholic numbness.

She passed in front of the Australian Embassy, which was located at number 15 of the same street, and then by the huge Finnish Embassy just next door at numbers 16 and 17. Embassy Row was becoming almost familiar to her now.

Moscow was dazzling; a resplendent bride all dressed in white, but there was no joy in the air, only a solemn, funereal-like sadness everywhere. She continued on her way towards the Metro station and her Russian language school. Russian was so difficult! She was struggling with the language but had promised herself to persevere. The lessons kept her busy and it would make Hassan so proud!

It was difficult to get around Moscow. She was always getting lost in the flat city. When she looked desperately for a landmark to guide her, she could only see the ugly Seven Sisters looming in the distance.

Those government skyscrapers, built during the Stalin period, were the only signposts on Moscow's outer circle. They were miles off and seemed to be always moving away from her, enveloped in the cold fog.

She walked for hours through the deserted, snow-covered streets looking for the entrance to the Metro. The dwarf was following her and pointing at her with a sneer on his distorted face. He was not wearing a heavy coat. She wandered around in the snow until she was frozen solid.

She cried for help but no one paid attention to her. Her voice seemed to be frozen inside her. She felt that she was about to faint from fright and cold, but miraculously, Hassan was suddenly at her side. He must have followed her. He tried to drag her towards a nearby mosque but she was stuck to the pavement. He started hitting her in an attempt to break the ice that enveloped her. How could he be so cruel? She screamed from the pain. An icy wind slapped her suddenly in the face and awakened her from her reverie.

She found herself at the entrance of the Metro; she must have been daydreaming. It took her a few seconds to collect her wits and remember where she was going. The Seven Sisters looked as menacing in the light of day as they did in her dream. It was a very cold November morning and by now the whole city would be frozen, and remain frozen for the next five months. Melanie shivered in her coat and decided to skip her lesson, and take the Metro towards Gorky Park.

The Metro station close to the embassy was a marvel. The wide, white marble stairs took her down to a palatial hall flanked with marble arches topped with high domes from which glittering crystal chandeliers were hanging. The lights were brilliant and the effect was dazzling.

The floors were covered with patterns of shiny marble. The station was like an elegant museum with exquisite sculptures alternating with carved marble panels. All that splendor was in shocking contrast to the ragged figures and the pale ghosts of people that were walking through the hall. The dull passengers distressed Melanie with their blank faces full of pain. She was thankful that it was only a short ride to the park and that she would soon escape from their oppressive looks.

Gorky Park was her favorite spot in Moscow. It was rather small, about two kilometers long and one kilometer wide. The language school where she took her Russian classes was close to its southern edge. She walked across the whole park easily, watching the children playing.

The Ferris wheel tempted her with its lights and its promise to lift her into the sky. She adored the view of the River Neves from high above, but today it was too cold to risk being suspended in mid-air, and she automatically headed towards one of the three skating rinks.

She had her skates with her and she realized that she must have decided beforehand to go to Gorky Park. Was she trying to fool herself? Skating was the closest thing to dancing and she yearned to stretch her body and flex her muscles. Hassan had forbidden her from skating without him and would be upset with her. It was not that he distrusted her, but he was so irrational about certain things. One of them was his fear that another man might touch her, even if it were inadvertently.

She would deal with him later, but now she'll skate.

She put on her skates; her heart racing in anticipation of the feeling she always got when she was floating on the ice and whirling around as free as a bird.

"Don't forget the cocktail party this evening," he had reminded her that morning for the third time.

"I won't forget." How could she? He always made sure that she remembered. He treated her like a child, trying to protect her from the world. She made a trial run around the rink and immediately felt reckless and free. She would love to dance and had wanted to take ballet classes, but Hassan told her it wasn't appropriate. He suggested that, instead of dancing, she could take classes in Russian.

Melanie smiled sadly to herself. Her eyes met the black eyes of a young skater. He smiled at her, approached her and held her by the waist. She let

139

him lead her in a dream-like dance to the music of Tchaikovsky. Her skating partner was just a boy, a toy soldier from the Nutcracker suite with shiny black hair, black eyes and pure alabaster skin. She felt a maternal tug in her heart. She shouldn't have agreed to marry Hassan; he was driving her crazy with his jealousy and control.

Maybe one day he'll loosen up as he slowly learns to trust her, she told herself.

The wife of the Egyptian Military attaché was there with her three children. Hoda was a pretty woman with a round face and a round figure. Her children were good skaters and Melanie sometimes skated with them while their mother watched from outside the rink bundled in her fur coat. The two womens' eyes met for a second and Melanie suddenly knew that she was doomed. Her secret escapade will be the subject of gossip and Hassan will find out. So what? She wasn't doing anything wrong. Yet, her heart sank with dread.

When she finally got home, Hassan was in his pajamas praying on the floor by the bed on the special prayer rug that his mother had given him. He was a million miles away, in a safe space that offered solace and promised heaven. She could never enter that space; it was closed to her forever and she was condemned to watch him from outside the circle.

Once in a while he would whisper aloud, *Allahu Akbar*, "God is great." This God of his may be merciful and forgiving, but he was not her God. He was a male God who set up men to be above women.

She even dethroned Buddha when she found out that he taught that a woman must cultivate her mind only so that she would be able to understand her husband and be able to help him in his work! Damn it!

All religions are male-centered.

Hassan's face was serene as he prayed softly. How she loved that man! He was so strong and yet so gentle. He said that he loved her, but how long will love last when he realized she would always be an unbeliever? He got up from the carpet, faced her, and spoke slowly in his deep voice, the voice of a biblical seer.

"You have betrayed my trust and I am deeply hurt."

"Hassan, I just went skating. It was a whim, not a betrayal."

"You were with another man."

"Oh, no, he was just a boy who happened to be skating. I don't even know his name. For heaven's sake, don't make an issue of an innocent pleasure."

"It is an issue. You have disgraced me and we can't go to the reception tonight. I just can't face anyone."

"Please, don't exaggerate. I was just skating."

"Melanie, you have deceived me."

"No, I would never deceive you. I just needed to dance, to move, to fly. Oh Hassan, please don't look at me this way. It was an innocent moment of pleasure."

"Mrs. Hosny didn't think so."

"The crazy bitch. I was just skating! Please forgive me. I'll not go skating again without you. I'll renounce skating altogether if that would make you happy."

"Melanie, if I'm going to spend my life with you, I have to be able to trust you."

"I've been honest and loyal to you. If you can't trust me then there is no hope for us."

"Maybe, I don't know."

"Bugger it! He's serious. It's ridiculous!" Melanie thought. She went into the living room and pretended to read for the rest of the evening.

Their life continued almost as if nothing had happened, but there was a coldness in Hassan's manner that rarely thawed. She caught him watching her closely, and she was positive that someone was following her when she went to her lessons. Maybe it was just her paranoia; it couldn't be true! He couldn't be spying on her!

They were in the bedroom and Hassan had a suitcase on the bed.

"Melanie, I have to go to Georgia, right away, for a few days."

"So suddenly? Can I come with you?"

"No, it's urgent and secret. Please help me pack a few things. I'm leaving in an hour. Will you be okay? You know that I worry about your safety."

She started helping him to pack. "I'll be lonely, but I'll be fine."

"An official from the embassy will come every day to check on you and see if you need anything."

"I don't need anyone to check on me," she laughed.

Hassan seemed to be uneasy, uncomfortable about something.

"What's wrong my love?"

"I'm going to lock the door. You can communicate through the opening at the top. You'll be okay for a few days." He went to the closet and brought out his blue suit, her favorite.

"Hassan, what are you saying? Are you crazy?"

He returned to the closet and brought out four shirts that she had ironed with a lot of love. He packed slowly with deliberate care.

"Oh, my God," she suddenly realized that he was really going to lock her up!

"You can't lock me up like a prisoner; I'm not your slave." Her voice was strangled with anger.

"Please, don't make it difficult for me. I've been thinking about it and I've decided that it's the best solution." He avoided looking at her.

"What solution? Are you serious? I'm to be your prisoner?" She decided at that moment that she wouldn't cry.

"No, you're my wife and I'm responsible for your safety." He continued to put things carefully in the suitcase.

"I'm a person, an adult; please, don't do this to me."

"Melanie, don't be so melodramatic. I'll be away for only a few days. The man will be bringing you some food and supplies."

"I'm going back to Egypt; you can't hold me prisoner. I won't stand for it!" Her anger was making her tremble; she felt trapped and automatically made a step towards the door as if she were about to run out. His eyes followed her calmly. He looked as if he was ready to pounce on her if she were to leave the room.

"You can't travel without my consent; besides, I have your passport."

"Hassan, please don't do this to me," she collapsed on the bed unable to control her fear. "I'm scared of being locked up. If you don't trust me, then let me go back to Egypt."

"We'll discuss all this when I return. I won't hold on to you by force. I have no time now. We'll talk about this later."

"What happened to you? Don't you know that I love you more than life itself?"

"Your definition of love is different from mine. Love to me, is honor, trust, and obedience."

"Hassan, I beg of you. You'd be destroying our love. You'd be destroying me."

She watched him as he shut the suitcase firmly, carried it off the bed and deposited it slowly on the floor.

"You're my wife. I'll protect you and protect my honor to the best of my ability. Five days won't kill you. If you love me, you'll obey my wishes and wait for me quietly."

The doorbell rang. She sat on the bed, turned to stone. Hassan hesitated a second, then left the room to open the front door.

Melanie heard voices but didn't move from her place. She hoped that she was going to wake up from this terrible nightmare. Hassan came in and picked up his suitcase. It wasn't a nightmare. He put down the suitcase again, came up to her and held her by the shoulders. He bent down and kissed her on the forehead.

"I'll be back as soon as possible; now be a good girl."

"Damn you to hell you stupid bastard!" She screamed at him silently.

Her head was about to split from a severe headache. She stayed where he left her, not moving for hours, until she either fainted or went to sleep.

She wasn't sure exactly what was happening to her.

The light of day woke her up. There were voices in the kitchen whispering. She slowly went to investigate but there was no one in sight.

Several paper bags full of groceries were on the kitchen table. She eyed them without interest and returned to her bedroom. She slowly closed the door on herself and took off her clothes. She put on her white silk pajamas and her thick white woolen robe, and then she crept back into bed under the covers. She cried for a long while until she was exhausted, and then she got out of bed and walked unsteadily to the white-tiled bathroom.

The large bathroom was cold and forbidding. The whiteness of the tiles reminded her of the tiles in a morgue. She imagined that she must be dead and lying on a slab with a number attached to her toe. Melanie looked curiously at her painted toes. No, she was standing up in this cavernous bathroom looking for her sleeping pills. She found them on the white marble shelf and quickly swallowed two pills with a glass of tap water. She put the bottle back on the gleaming white shelf, but then reconsidered and decided to take the bottle and the refilled glass of water to bed with her.

She knew that she was drugged because she was having visions of herself sinking in the bed as if she were drowning in the very soft and clinging mattress. She could hardly breathe and felt that she was on the brink of suffocating. Images of her life flashed across her mind like a speeded up film.

Time seemed to have stopped; whenever she woke up, she took a couple of sleeping pills and went back to sleep. She lost count of how many times and how many pills. She heard a voice calling her; she got up in the middle of the night and dragged herself to the kitchen. There was no one there. She was almost crazed with fear but forced herself to drink a glass of milk. When she returned to her bedroom, she found the dwarf sitting on the bed.

"You're pathetic. How're you going to escape from this prison?" he said.

"It's not a prison; it's my home."

"You never had a home silly girl."

"Leave me alone you ugly monster!" She started screaming at the dwarf who slowly moved away and left the room. She chased after him but he was nowhere to be found.

The vision vanished as suddenly as it had appeared, but left her shaking with dread. Melanie cried out and screamed for help but no one seemed to hear her. Finally exhausted, she lay down in bed and swallowed two more sleeping pills.

The apartment was getting cold because she forgot to put coins in the heating meter. Unaware of the passing of the days, she barely ate and hardly moved from the bed. As her strength vanished, her sense of reality evaporated. She thought that she heard a loud knocking on the door, but she was unable to get up to find out who it was and besides, she didn't care. She remained huddled under the covers sweating from fear.

She swallowed more pills and was swiftly carried away into oblivion.

She was finally happy, free of pain, but someone was shaking her and keeping her from finding peace.

"Melanie, wake up. Please, my darling, talk to me."

She must be dreaming. Hassan was holding her in his arms and kissing her. All she wanted to do was to sleep; she didn't want to dream anymore.

"You're going to be all right. Please, drink this soup."

"She's running a very high fever," a voice said.

It sounded like her friend, the dwarf. Why is he talking to Hassan? She tried to warn him but was enveloped in darkness again. When she regained consciousness, she was in a hospital bed with tubes attached to her body and a heavy weight on her chest. She could hardly breathe.

"Hello there! You really scared us young lady; you actually have pneumonia, but you'll be alright." Hassan was bending over her with his hand over her forehead.

"You came back," she murmured in a distant voice.

"I'm here and I'll never leave you again, I promise."

Melanie got well slowly while the Moscow spring was thawing the snow, but she never really fully recovered. She was now afraid to go out alone and stayed close to Hassan whenever possible. He took care of her as if she were a helpless child, and made an effort to be constantly by her side.

She received some letters from Nadia with pictures of the baby.

Melanie was shocked when she discovered that the baby was nearly a year old.

"You're a slave. How're you going to get out of this prison?" The words were ever present in her foggy mind.

SUEZ CANAL WAR

T HE EVENING OF SEPTEMBER 10, 1956 was clear and cool. The stars shone in the sky like a jeweled veil covering the dark face of an Egyptian princess. On the ground, Sammy and the regiment to which he was assigned were moving in a loose formation across fields freshly plowed and planted with wheat. The earth was soft and black. The rich Nile silt was inexorably feeding the plants, untroubled by the stomping of hundreds of ill-fitting boots and antiquated heavy equipment. Some of the soldiers, who were recently recruited peasant youths, took off their boots and walked barefoot on the worm-infested soft earth. The *Belharzia* worms, invisible to the naked eye, would travel though their skin and find a permanent home in their young livers. The disease was endemic in the Nile valley and caused untold damage to the people's health.

The regiment was supposed to be creeping stealthily up on the British and French camps in Port Said, but the Egyptian soldiers were very raucous; they sang and joked while marching behind the irregular lines of tanks and jeeps. They appeared to be schoolchildren going on an outing, not soldiers going to war.

Sammy rode in a jeep, behind one of the medical ambulances, with his partner Dr. Hisham Zaky. Hisham was tall and swarthy; his black velvet eyes permanently hidden behind thick spectacles and his thick lips in an eternal pout. Sammy called him the gentle giant.

Sammy and Hisham had previously done their perfunctory military training together and were now captains in the medical corps.

"In the army you have to abandon your reason completely and follow orders blindly. It goes against my deepest instincts," Sammy told his companion.

"I like to avoid trouble at any cost. All I want is to be a doctor; I didn't bargain on being caught up in this mess, but there is no point in rebelling," Hisham answered.

Like Sammy, he found himself in the army out of expediency.

Sammy's eclectic interests bewildered Hisham but he admired his partner's daring and original thinking. In spite of their different temperaments, they got along well together.

"Is it true that they destroyed all our aircraft and runways?" asked Sammy. "It seems incredible."

"I'm almost sure, the BBC is pretty reliable," said Hisham in his baritone voice. He was always the voice of doom and gloom, and now he was in his element, thought Sammy.

The two young doctors were part of the military convoy moving slowly from Zagazig to Port Said. Their mission was to surprise the French and British armies that had occupied the city.

"I can hardly stand listening to the lies on Radio Cairo," Sammy sputtered in anger.

"The fact is, my dear friend, the British, the French and the Israelis have parachuted safely into the Canal Zone and Sinai after they paralyzed our air defenses," said Hisham.

Their attention was suddenly diverted to the loud noise of a crash.

At an intersection in the road, a cart carrying live chickens in bamboo cages had crashed into a military jeep and had overturned. The chickens flew out all over the convoy. A low-flying Israeli biplane materialized at that moment and started shooting at the convoy. The co-pilot had a hand-held automatic machine gun. Chaos broke out, cars swerved, and soldiers took cover in the irrigation ditches.

The peasants, in the mean time, were oblivious to the gunfire and were running around collecting the chickens that were scattered like noisy children let out of school. The plane finally seemed to lose interest. Sammy could see the pilot laughing as he flew away. The convoy continued its march.

Due to the war, all of Sammy's plans to leave the army were put on hold. His moviemaking career, which wasn't doing that great anyway, had come to a full stop. The only issue now was to survive the bloody war. Sammy wondered what Nadia would do if he were killed. He missed her and his little boy very much. Sammy was angry at being trapped in what he considered a futile war. His mind didn't stop churning. He longed for his son who wasn't even a year old and for Nadia. "What am I doing here?" He asked himself, "I'm not supposed to carry a gun; I'm not fighting. I'm here to save lives not to kill."

"There is no reason to be here. The war is over and we're occupied again," he said to Hisham.

"I wish Eisenhower hadn't refused to finance the Aswan Dam project; we wouldn't be in this mess," answered Hisham.

"A wish . . . , my old nurse used to say . . . can leave you as hungry as an empty dish," said Sammy smiling.

"I knew the minute Nasser nationalized the canal," said Hisham, "that the allies would occupy us as they occupied us a hundred years ago. The Suez Canal is too strategic. It's been a curse on Egypt from day one."

"It's the usual short-sighted American policy; for only seventy million dollars they threw us into the arms of the Russians." said Sammy.

"I just can't believe that we'll be occupied again," continued Hisham, his eyes watering behind the thick lenses.

"It's not so hard to believe. Do you remember the bus stopping on the way to Zagazig and the driver announcing loud and clear to all the passengers, 'The Secret Airport, this is the stop for the Secret Airport", Sammy laughed. "That is how we prepare ourselves to defend this country, secret airport, my ass."

"No wonder it was so easy to target our miserable Air Force," answered Hisham.

"Our planes were like a herd of sitting camels resting in the desert.

They destroyed all of our airports in one day, without meeting with any resistance from the mighty Egyptian army," said Sammy.

"Sammy, it's just suicide to fight the Israelis in the desert without any air cover."

"We're lucky that we're in the National Guard and have only Zagazig to protect," said Sammy. "I don't want to die because of the mistakes of the Generals. If I survive this war, I'm quitting."

The two men continued their conversation from Zagazig to Port Said. Sammy watched sleepily as the peasants came running across the fields, their lighted Kerosene lamps flashing in the dark like an attack of colossal fireflies. They called out to the soldiers with words of encouragement and praise. It was a golden opportunity to sell the army fresh fruit, candy, roasted corn, and even chicken.

Suddenly they heard shots fired in the air. The British soldiers, who saw the Egyptians coming closer, tried to warn them not to proceed any further. The Egyptian major at the head of the line was surprised to find himself almost upon the British camp. He was too embarrassed to retreat and ordered his soldiers to set up camp two hundred feet away from the barbed wire that extended around the enemy's camp.

"We're much too close," said Hisham, his face showing frustration and fear.

"The British must be laughing at our stupidity," said Sammy, "They could've slaughtered us easily but took pity on our pathetic situation.

Maybe we're not worth the effort." Sammy wasn't afraid, but only felt a deep, helpless anger.

"I guess they didn't want to waste their ammunition. They've already won," said Hisham bitterly. "What kind of a war is this?"

"A stupid war, like all wars. Actually, when you come to think of it, it's not too bad," said Sammy, "I'd say it's very civilized. Just like medieval battles when enemies faced each other and fought an honorable fight."

"Honorable fight? That's a contradiction in terms," murmured Hisham.

"*Shay*, tea?" asked a young peasant approaching them smiling. He had recognized Sammy from one of his movies.

"*Shukran*. Is there anything to eat?" asked Sammy smiling.

"Certainly, *Sidy*, just one second, master."

The Egyptians lit bonfires around which they gathered. The smell of hashish began spreading; then the smell of roasted meat started mixing with the hashish. Sammy could hear the tinkling of the spoons mixing sugar into the glasses of boiled tea, and now and then, a laugh burst out like a gunshot and startled him.

They found themselves joining a group of officers around one of the bonfires. The peasants had brought a slaughtered lamb as a gift to the valiant officers and someone started to barbecue the meat. Freshly baked pita-bread, feta cheese and green onions materialized. Cold beer and a starlit night turned the evening into a feast.

In spite of the pretense at festivity, the officers were tense. The jokes were interspersed with bitter and cynical remarks about their plight.

"Every time we try to raise our heads or have a leader who thinks of what's good for the country, the superpowers intervene and crush us like bugs," said Hisham.

"We deserve whatever we get," said Sammy, "We're spineless."

"Sammy, you are so full of self-hate. Don't be so cynical." Hisham was worried that Sammy might antagonize the other officers.

"The Egyptians have always survived because they are strong and yet malleable; we bend with the wind," answered a Lieutenant with fiery eyes.

"Egyptians are not fighters; we have always been servile," countered Sammy. "Just look at our history, we've always been submissive slaves since the time of the Pharaohs."

"We're all ready to die for our country. Who are you calling slaves?" said the angry Lieutenant.

Hisham forcefully pulled Sammy up to his feet. They'd been sitting on the ground like boy scouts around the campfire.

"You're tired, my friend; let's go get some sleep."

Sammy got up with a sigh and wearily saluted his comrades.

"I'm sorry if I offended anyone, *Tesbahu ala Kheir*. I'm just angry and disturbed. May you wake up to a prosperous morning."

The farewell greeting sounded ridiculous to Sammy and out of place. He doubted if anyone else realized its incongruity.

"Doctor, my advice to you is to get yourself a gun. Tomorrow might be a less peaceful day," commented the fiery-eyed Lieutenant.

"I don't want a gun; I'm a medic, not a soldier."

On their way back to their quarters, Sammy noted how the British camp was dark and silent just within sight of the festive Egyptian camp.

"I'll sleep in the back of the ambulance. It's safer," he said.

"You think that they won't bomb an ambulance?" laughed Hisham.

"Who mentioned bombs? I'm worried about the mosquitoes. They just love my tender skin."

"*Tesbah ala kheir,*" Hisham called back, already on his way to his own tent. Sammy went to the ambulance and made himself comfortable on one of the stretchers. He cursed under his breath at the bitter cold and piled blankets on top of himself. He tried to sleep, but the carefree sounds of the young soldiers who were exchanging jokes and chatting in loud joyful voices around him, kept him awake. Images of cut off limbs flying through the air, soldiers without any legs begging to be killed, faceless boys whimpering, and mutilated children staring helplessly haunted him. The casualties arriving at the hospital in Zagazig were still fresh in his mind. His last thought, before he finally fell asleep, was that they were all like a flock of dumb sheep marching off a cliff to their certain death.

Finally, an orange dawn started creeping up the sky, and a brilliant sun began warming up the fields. The colors and smells of the early morning woke the men sleeping in the trenches. Hisham opened the door of the ambulance slowly and looked in.

"I got you some tea; the major wants to see you, "he boomed.

"It's not even six o'clock! What does the asshole want?"

"I think we have some problems, but I'm not sure what they are."

"Great, I just hope that he's got diarrhea from eating too much," said Sammy as he got up, and quickly prepared himself to meet the fearless leader.

"We need you to negotiate the passage of a train carrying the dead and wounded soldiers from Port Said to Cairo," said the Major. "I was told that your English is good."

"Yes sir, I'm a regular product of the British occupation. My English is better than my Arabic."

"I wouldn't brag about it if I were you. Anyway, this is the plan.

You'll take two unarmed soldiers carrying white flags with you and go over to the senior officer at the British camp. Ask him for written permission for the train carrying the dead and the wounded from Port Said to pass through. It'll be under the auspices of the Red Crescent."

"What if they start shooting at us, sir?" asked Sammy.

"Well, if they shoot, we have you covered. We'll shoot at them," said the Major.

"What good will it do me? I'll be caught between the two fires."

"Sorry, that can't be helped," replied the officer with a straight face.

"Thank you very much sir," said Sammy, and went out to get two volunteers for the suicide mission. "We have you covered," Sammy cursed under his breath. "Damn the religion of the midwife who pulled him out of his mother's cunt," Sammy murmured to himself. He hoped that the British would be wise enough not to shoot at him.

He carefully put a white band with a red crescent on his arm and on those of the two soldiers who had volunteered to go with him. They held up two white flags and off they went. The three marched solemnly towards the British camp. The soil seemed to be alive; it was moving in the hot morning haze. They could see a British soldier stripped to the waist already turning as red as an *Amhat* date ripening in the burning sun. Noel Coward's song, 'Only mad dogs and Englishmen go out in the Midday Sun,' kept running in Sammy's head.

"What do you want?" the soldier barked at them.

Sammy's heart was pounding like a brass pestle pounding kobeba meat. What on earth was he doing? He was still a boy scout risking his life for a good cause.

"Sir, I'm here to ask permission for the Red Crescent train to pass from Port Said to Cairo with the dead and injured." Sammy said.

"Wait here. I'll get you the Major."

Sammy turned to signal to his regiment that everything was okay.

He was more afraid of them than of the British. Someone might panic and decide to shoot for no reason.

A red faced senior officer soon came up to the barbed wire. Sammy saluted him and the British officer returned the salute.

"How dare you show your face and ask us for a favor. You're nothing but animals. Go back and tell your chief officer that not only will there be no train passing, but that we'll do to him what you buggers did to the two journalists you caught last night."

"I'm sorry, sir, I have no idea what you're talking about."

"For your information, your bloody people took hold of a press jeep that had lost its way. They not only massacred the poor blighters but they mutilated their bodies in a savage way. Then they sent them back to us in an unspeakable condition. Now they want the train to pass?

They're not only savage, but stupid as well."

"Yes sir, I totally agree, but fear combined with ignorance can make men lose their minds. The culprits must have been peasants venting their fury on the enemy."

"Are you a doctor?" asked the officer.

"Yes, sir."

"You speak English well."

"I lived in England for many years before the war, sir, and then I went to an English school in Cairo. I'm what you call a WOG, sir."

Sammy smiled his crooked smile.

"A western-oriented gentleman! I haven't heard that expression in years," the officer laughed.

"Laugh you son of a bitch," thought Sammy, "your arrogance and your Empire are doomed." The soldiers accompanying Sammy suddenly relaxed and smiled, even though they didn't understand a word of what was said.

"What do you say, sir, let the train pass? It's full of wounded soldiers. The powerful can afford to be generous."

"Okay, we'll let it pass, I'll send a message. Tell your chief that there is outrage about the two reporters; we'll have to have the Red Cross investigate. The criminals must be brought to justice."

"Yes, sir, I understand. Thank you, sir."

Sammy saluted and turned around walking fast towards his camp.

He felt proud that he had succeeded on his mission but wanted to strangle his commander, son of a dog, who never told him about the killings.

The next day, Sammy was on the platform of Zagazig's railway station waiting with his medical team for the train full of casualties to arrive. The train came chugging towards the station and stopped. Some of the volunteers got off to meet the press and stretch their legs. Sammy heard a shrill voice calling loudly from the train steps. It was a pretty woman in a tight Red Crescent uniform, and she was yelling as loud as she could, "*Sousou, Sousou, habib alby*, love of my heart."

Sammy recognized one of his relatives. To his horror he realized that she was calling him. He pretended not to hear her and turned his back to the train, but she wasn't discouraged. The adrenaline was flowing in her veins from a mixture of fear and relief. She persisted.

"*Sousou*, habiby, my love." She was now running towards him and at this moment he would have liked a genie to carry him off and place him on top of a mountain.

"*Sousou*, what brings you here?" She cried as she hugged and kissed him in front of all the men. He was now assured of being the butt of jokes for the rest of his life. The soldiers were enjoying the show.

"Ahlan, welcome Auntie Nawal," Sammy said. You stupid cow, can't you see I'm with my soldiers performing military duty? Sousou indeed!

Sammy would have liked to strangle her.

Sammy remembered how this same women was a guest at his home and had slept in the same bed with him when he was about eighteen.

He had become so aroused that he had ejaculated on her thigh while she pretended, all the while, to be asleep.

"Your mother will be so happy. I'll tell her that I saw you."

"Please do, many thanks, Auntie, and please, stop calling me *Sousou*," said Sammy.

"And you stop calling me *Auntie*. A grown man calling me Auntie! I'm not much older than you, you know."

"And you are very lovely too. I'm sorry, it's just a matter of habit.

You'd better get back on your train. You've caused enough trouble already. *Ma'a El Salama*, go with peace."

She left and he turned back to his men.

"I'll shoot anyone who calls me *Sousou*. Is it understood?"

"Yes, Doctor *Sousou*," a voice answered quickly.

Sammy looked at the young faces sternly, but then, he couldn't help laughing out loud himself.

The war was over, the fighting had stopped, and the politicians had started talking. President Eisenhower forced the British, the French and the Israelis to retreat, and Nasser declared victory. The army claimed victory, but the people, who were licking their wounds and burying their dead, knew that there was nothing to celebrate.

At home Sammy amused everyone around him with tales of his adventures during the short war. He kept the ugly stories to himself. In spite of his happy façade, the war had changed Sammy and hardened his resolve to quit the army for good.

Sammy had barely returned to his previous routine of hospital and clinic, when he was suddenly posted to Sinai and found himself sitting in a makeshift dispensary, in the intolerable summer heat, looking out the window at the vast desert facing him, like a dead colorless ocean.

"They hate me in the army, and this post is a punishment. I have no protection now that my father has retired. They want to make my life miserable," he said.

"Why should anyone hate you?" Nadia said.

"Nadia, they don't approve of my acting and they don't like my record in the army. They suspect me of being too independent and they are right. I have to find a way out."

One morning Sammy saw a mangy stray dog approaching him.

The animal was lurking around, obviously looking for water and food.

A brilliant idea occurred to him. He got up slowly and shut the door, then he took a pair of tweezers from the surgical supply tray and began to squeeze the skin of his arm into what soon looked like a realistic dog bite. He fed the dog a couple of biscuits and then shoed him off. When the dog was far away from the camp, Sammy sent out a blood-curdling scream. "He bit me! He bit me!" he cried.

"What happened?" The hospital staff came rushing in.

"You better catch that dog. I think he must have rabies." No one was very eager to chase a mad dog in the burning desert sun. The animal disappeared. Sammy pretended that it was no big deal, but the senior officer insisted on sending him to Cairo on sick leave. He had to immediately get ten weekly rabies shots and rest until the danger of being infected was over.

It worked out just as he had planned. Sammy used his sick leave to finish his movie, *Cousin Fatma*. The film was a great success but he was criticized as being pro-western. The Minister of Culture called him in and told him that they would give him work, but that in the future they would select the subject of his movies.

"We want films with high moral aims, films that reflect our national aspirations and the philosophy of the revolution."

The situation in Egypt was deteriorating rapidly. The secret police, modeled on the Soviet KGB, was now blatant in the abuse of their powers. It was rumored that some actresses were being forced to have sex with high officials. The actress, Samia Rashed had jumped from her window in order to escape being raped. She was lucky; she lived on the ground floor. Merchants were not being paid for their goods and were threatened with jail if they complained. A puppet parliament was established; they saw on television a dissident being carried away physically from the meeting hall and thrown into the street. All political parties were abolished. The Moslem brotherhood, a banned religious organization that opposed the military, went underground and gained in strength.

When the government nationalized the movie industry, Sammy decided it was time to leave the country. He jumped at the opportunity when a Lebanese producer invited him to come to Beirut to establish a new movie industry.

He presented his resignation from the army and waited impatiently for his dismissal. During the long wait, Sammy contrived in the most imaginative ways to get prolonged leaves of absence. He faked an appendicitis attack and even went so far as to undergo an appendectomy in order to stay home waiting for his discharge.

"*Dear Melanie,*" Nadia wrote. "*He plans to go to Beirut to make movies.*

Though it is only two hours away, Beirut could be on the moon. It's almost impossible to get exit visas nowadays. I might not be able to see my parents for many months. What should I do? Most importantly, I will have to stop my studies towards a masters degree. When am I going to see you again? I've seen you only two times in the last three years. I need you. I love you. Nadia"

There was no answer to her letter; Nadia felt abandoned.

21

DEATH IN THE FAMILY

S HORTLY AFTER THE END of the Suez war, during dinner at his in-laws, Sammy announced to one and all that he was leaving for Lebanon with his wife and son. Omar Soliman turned red and the large vein visible on the left side of his forehead started pulsating.

"If you want to go to Lebanon, go alone and see how things are before dragging your wife and son into an unknown situation," he said.

Omar threw his napkin down and stood up suddenly, almost knocking over his seat. He stood for the longest time with both hands on the white lace tablecloth, staring at Sammy with pent up rage.

"What are you planning to do? Will you become an actor, a director, a doctor, or a writer?" Omar's voice was raised to the level of yelling.

"He hates my guts," thought Sammy, smiling amiably at his father-in-law.

"I've been offered a very generous contract to produce movies in Beirut," Sammy told him.

"Now a producer!" yelled Omar. "May God protect us."

He sat down for a moment and continued eating slowly. "This food is tasteless," he told his wife. "Tell Osta Taha to come here right now,"
he ordered the butler.

Nadia was glad her son was not witnessing this outrageous scene.

They all continued to eat in silence until the cook arrived from the kitchen. Omar Soliman suddenly stood up again and threw the silver dish containing the eggplant *Moussaka* at the cook's head.

"When are you going to learn how to cook?" Omar shouted at the red-faced cook. "The eggplant is raw and bitter and the sauce is watery.

One more dish like this one and I'll send you back to the army to peel potatoes."

"I'm sorry Omar Bey, I'll be more careful next time, master." Osta Taha left the room with his head dripping red tomato sauce and with pieces of thin sliced eggplant hanging on to his white uniform. His ego was as bruised as his aching head. He was a chef, not a cook's boy, yet he accepted the insult meekly. He needed the job.

Nadia was mortified. She hated the way her father treated the servants. Taha was the same age as her father and he had been their cook since before she was born. She used to love sneaking down into the kitchen to watch him cook, and he would always have little treats to give her.

When Sammy spoke again, his voice was still shaky but almost normal.

"I'm sorry, Sir, but I want my family with me. We'll rough it together. Don't you agree, Nadia?" Sammy gave her a look that came out much more like a command than a supplication; he wanted her to declare her willingness to go. He believed that she should uphold him in his decision even if she had stated to him, in private, her doubts about his plans. He was beginning to feel that sending her to college was not such a good idea. She was having too many ideas of her own.

Nadia didn't answer. The idea of leaving her parents and Egypt was very frightening, but on the other hand, she couldn't even think of leaving Sammy. She had to stand behind him whatever he decided to do. She was caught between her father and her husband. She knew that she was being a coward, but still, she didn't speak up. Her mother started to cry.

"How can you be so cruel?" Amina asked, looking at Sammy. "How can you deprive us of our grandson after he's become a shred of our hearts?"

"Keep quiet, Amina," scolded Omar. "Sammy is right. He has to make a break from the mess he created here, juggling a medical and a cinema career."

Amina left the room quickly, and Nadia bit her lower lip in order to control her tears.

A few days later at the Hakims' home, Sammy told his parents that he had decided to go to Beirut to make films.

"I'm not surprised," Dr. Hakim said quietly. "I knew you'd never become a real doctor."

Dr. El Hakim never had any illusions about his son's disinclination towards a medical career. He also knew that Sammy had serious problems with the army. Among other minor infractions, he had sold a gun issued to him by the army.

"Maybe it would be best for you to leave Egypt at the moment and try to make a fresh start," he said.

"What are you going to do in Lebanon? You don't even know how to cook." Laila told Nadia.

"I'll have to learn, I guess," said Nadia looking at Sammy for help.

"We won't starve." Sammy said.

"How about money? How are you going to afford to move?"

"We're going to sell everything."

"Are you completely crazy?" said Laila, "the furniture belongs to your wife. You can't touch it."

"It belongs to both of us and won't do us any good sitting in storage. Besides, we're never coming back."

"I can't believe what you're saying," said Laila. "What kind of craziness is this?"

"Mama, please," Sammy told her. "Let's not discuss this anymore. I'm going to Beirut. I'm a grown man and I think I need to get away from here and focus on a life plan."

"Life Plan? What crazy life plan is that? You mean to throw away all the efforts we made to get you a medical degree?"

"I'm the one who stayed up nights and ate dog meat for seven years to earn the useless degree. All I gained are flea bites and crippling debts."

There was a pained look on Dr. Hakim's face.

"You're going to desert us and go to Beirut after some kind of Hollywood dream?" said Laila. Laila's face turned a dark red. She, who controlled the lives of all the people around her, was thinking of how bitter it was that her adored son was trying to escape. She felt dizzy and knew that her already high blood pressure must be out of control.

Dr. Hakim sat quietly eating his meal.

"Omar Soliman is right; you're a *mostahter*, reckless and wild."

Laila said. She knew very well how much Sammy hated the epithet of *mostahter*. She ignored Sammy's furious looks and turned to Nadia, who was sitting quietly with her eyes cast down, pretending to examine the intricate design on the carpet with deep interest.

"You, Nadia, can't you speak up? It can't be what you want. Talk to your husband. Life in Beirut won't be easy. You'll be alone with the boy. Let me assure you, you'll rue the day."

"I'm leaving whether she comes with me or not." Sammy told Laila.

"My mind is made up."

"You should consider the comfort of your wife and son," insisted Laila.

The battle for freedom was between him and his mother. Laila represented all the shackles that tied him down.

"In this family," he said pointing to Nadia and himself, "I'm the boss."

Dr. El Hakim lifted his head from his plate and looked at his son.

He didn't utter a word but Sammy was mortified; he knew that he had offended his father.

Nadia was taken aback by Sammy's response. She wondered if he would really leave her behind if she voiced any objections?

For weeks after that, while they were preparing their papers to leave, Laila argued daily with her son. When direct attack failed, she changed her tactics. She cried and begged him to stay.

"Please, *ya habibi*, my love, don't leave me. I can't live without you."

She cried every time she saw him.

Sammy finally lost his patience with his mother.

"Mother, can't you understand I hate it here? I'm not sure if I'm suffocating from the military rule or from you. I'm leaving for good and never coming back. I want to get away from your choking embrace.

I'm not your baby anymore, neither am I your lover. Just let me go. Leave me alone."

"How dare you talk to me like that!" Laila said, "Don't you know that the prophet said that heaven is under a mother's feet?"

"He never said that you have to crush your children under those feet," said Sammy.

The day before their departure, Nadia and Sammy, with their son, went to his parents' for a farewell dinner. Laila had prepared a sumptuous feast.

As was usual in the home of Dr. El Hakim, there were many self-invited relatives sitting around the dining room table. Nadia counted twenty heads including the children. The table was covered with enough food for fifty.

"I think Auntie Laila," said Nadia, "you have cooked all of Sammy's favorite dishes."

"I'm sure you'll never be able to cook Egyptian meals like these for him," said Laila.

"No, Mother," said Sammy. "We'll only be able to dream about your cooking."

"I'm glad you'll miss something," said Laila.

"I heard that you had been called in for questioning at the military headquarters. What was the problem?" said his cousin Ali, his bushy mustache quivering. Ali was obviously in the not so secret service. He knew too much, thought Sammy.

"Nothing important," Sammy told him. "They were asking me how on earth I could be acting in the movies while I was supposed to be a doctor in the army," said Sammy.

"I always wondered about that myself." Ali's voice, which was as mellifluous as molasses, irritated Sammy.

"I showed them a form signed by General Abd El Hakim Amer himself and that stopped any further questions," replied Sammy.

Warda, Ali's wife, left the room in order to get something from the kitchen. Sammy tried not to stare at her heavy hypnotic hips as she walked away.

Six years ago, one sleepless night, Ali had raped his pretty cousin Warda, who had been visiting at his family's home. When the girl, who was an orphan, ran to Sammy's mother for help, Laila took control of the situation and forced Ali to marry the girl. Ali and Warda never really liked each other, and their marriage was notoriously turbulent, but they managed to produce one retarded son. Sammy was sure that Ali stayed married to Warda only out of fear of Laila.

"How on earth did you manage to get General Abd El Hakim Amer to sign such a form?" asked Ali.

"Easy, I approached the General," said Sammy after a moment of silence, "when he was in a good mood and high on hashish."

"Be careful what you say," said Ali, looking around him.

"Anyway, there aren't any more problems with the army; my resignation was accepted, as I'm sure you know, and I have a work visa from the Lebanese Embassy."

"Why are you choosing to exile yourself in Lebanon? I don't understand. Don't you know that Egypt is the mother of the world?"

said Zohra, the obese wife of Sammy's uncle, Hamed El Hakim. Her hips were so huge they formed what Sammy had dubbed, *Rafarif* or fenders, around her huge body. Zohra had come over with her family from the village of Shibeen to visit. They had brought baskets full of vegetables, fruit, baked bread, and cookies. Sammy didn't answer the woman. He pretended to be busy eating.

The guests were occupied with gorging themselves on the food and didn't notice that Laila had left the table. She stood at the door of the dining room looking at her son with sad accusing eyes and suddenly lost her balance; she fell down in a heap on the wooden floor. Everyone stood up in panic. Doctor El Hakim and Sammy tried desperately to revive her. They staggered with her heavy body, carrying her the short distance to her bed in the next room. The two doctors, father and son, attempted to resuscitate her, but failed. Dr. El Hakim felt helpless and inadequate. He, who had saved so many lives, couldn't save the person he loved most. Help came swiftly from the nearby hospital, but Laila had fallen into a coma from which she would never recover.

Nadia fixed her gaze on her mother-in-law. She sat in shock by her bed all night watching every move and listening to the rattle of Laila's breath. The amazing clutter in the bedroom made it look rather small.

In fact, the room was large with huge windows on both sides of the bed. The windows were covered with heavy satin drapes that seemed to have been collecting years of desert dust. The sheer curtains under the drapes were a muddy brown color instead of their original off-white. On top of the armoire that stood in front of the bed against the wall, were woolen blankets and satin eiderdowns, folded carefully and stacked for use against the short, but bitter-cold Egyptian winter.

The art deco dressing table in the corner was piled with empty perfume bottles, half-used cream jars and a jumble of mostly expired prescription medicines. Fighting for space on the dressing table were countless family pictures that dated over a period of sixty years. Among all these objects there were dozens of magazines, newspapers, and a few unopened bills and letters.

Many Doctors filed into the chaotic room and consulted with Dr. El Hakim, who kept going in and out of the room in a daze. "Why are you leaving me now when I need you so much?" Nadia heard him whisper to his dying wife. It broke Nadia's heart to see the pain on his face. She sat there paralyzed, feeling numb and helpless.

When, before daybreak, Laila stopped breathing, Nadia cried out in anger. She felt that Laila had died just to punish her and Sammy for daring to leave her. Later, grief would overcome all other feelings.

Sammy was devastated. Nadia found him later, standing alone on the balcony outside his old bedroom. He was crying inconsolably. "I murdered her," he wailed. "I might as well have shot her with a gun."

"What are you talking about? She had high blood pressure and she suffered a stroke; you're not responsible. Be reasonable, darling."

"I killed her. I told her I was never coming back and it killed her."

"You can't feel guilty, my love. It wasn't your fault."

"Now we have to change plans," said Sammy, "How can we leave now?"

They would have to postpone their departure. They would have to find a new place to live in since they had relinquished the lease on their apartment. What a disaster!

"Laila got her way," said Nadia, "she has stopped us from leaving."

The women in the room had started wailing and screaming until Zohra, the expert on sudden death, calmed them down.

"*La Ilaha Illa Allah*; there is no God but Allah. It is the will of God.

We must accept the will of God like good Moslems. Let us pray so that her soul may rest in peace." With that, Zohra started reciting the first verse of the Koran:

> *In the name of God the merciful, the benevolent.*
> *Thanks be to God, the God of all creation.*
> *The merciful, the compassionate.*
> *Lord of the day–of–judgment . . .*

Everybody left the room except for Zohra, who was praying, Nadia, who was trembling with fright and couldn't move, and Warda, who was crying and talking to the dead woman.

At the break of day, a Sheik with a beautiful voice was heard chanting the Koran in the living room. The sound of the Koran quieted all the women except Warda, who continued with her lamentations.

"Why have you left us? It's not your time to go, my precious." Warda kept on with her one sided dialogue with the dead woman until a knock was heard at the door. Four hefty women entered Laila's bedroom with a portable plastic-covered table. The black clad women, who were sent by the mortician, were there to wash the body and assist in preparing it for burial. They undressed Laila and put the naked body on the table. It was no longer a person, but a slab of meat. To her horror, Nadia noticed that the blood was beginning to settle in the lower part of the body.

The women started to wash the body with soap and water while they murmured verses from the Koran. They washed her eyes, her nose, ears, and all her orifices. Nadia watched, mesmerized, unable to leave the room or take her eyes off the body. The women took a break in the middle of their job to drink tea and smoke a cigarette. When they finished washing the body, they dried it very well. They stopped and one of them went up to Zohra, who was standing in a corner of the room looking slightly deranged. She whispered something in her ear and Zohra promptly fished in her pockets and put some money on the body. One of the women picked the money up and put it in her bra. It was the custom that the dead person should be the one who pays the washers. After they took the money, the women started to wrap the body in a white cotton shroud. They then wrapped her in one of the new satin eiderdowns.

All of a sudden Sammy barged into the room that was now filled with wailing women. He started screaming at the women in a voice that chilled Nadia to the bone.

"Stop screaming at once! Stop this charade! I don't want to hear any screaming!"

There was a sudden silence, and then the women started weeping quietly. The wrapped up body was carried out of the bedroom by four men who lifted it on a wooden palette and proceeded to walk down the stairs, six floors down. The crying turned to wailing, and soon the women were screaming again at the top of their lungs. Some of the mourners were slapping their faces, pulling their hair, and tearing up their clothes. The scene was just like the ancient Egyptian mourning scenes Nadia had seen on the walls of some of the Royal tombs in Luxor.

Sammy accompanied the body to the entrance of the building where the male mourners were gathered. The body was placed in a simple wooden box and covered by a silk paisley cloth, and the men of the family escorted it to the mosque that stood at the end of Sesostris Street. There were hundreds of mourners praying at the mosque for the soul of Laila.

For the first time in a cemetery, Nadia watched with feelings of pain and horror, as the body wrapped in the cotton shroud was laid directly on the dirt in an underground chamber destined for the women of El Hakim's family. Even in death, the females were segregated from the males. There was nothing pretty about the burial; it was simple and final. No flowers were put on the grave and no head stone was set to mark the spot. Nadia wished she were a devout Moslem, a person who accepted without question the will of God, for it was for moments like these that religion was invented.

Later, Sammy and Nadia went home to sleep on the floor of their empty apartment. Ahab had been taken to Nadia's mother.

"What are we going to do?" asked Nadia with tears running down her face. "We can't leave your father alone under the circumstances."

"My father insists that we don't change our plans," said Sammy.

"But it isn't right;" said Nadia, "we have to wait at least the forty days of mourning."

"I'm confused; I cannot think," answered Sammy.

Nadia felt numb with despair. She knew that Sammy was stunned with guilt and grief and she tried to console him. She held him in her arms and caressed him while he sobbed.

"I'm a murderer," Sammy kept on repeating.

"She killed herself with all that rage. You have the right to your own life. Don't blame yourself. Please be reasonable."

"I'm worthless."

"You're not worthless," cried Nadia, "How could I love a worthless man? You're going to be a great success. I know, I'm sure of it." She kissed him tenderly.

Nadia was surprised when she realized that he was getting aroused and that he suddenly wanted to make love to her. Sammy made love to her with a fury and a passion that dismayed her. She felt that making love was somehow indecent under the circumstances, but she kept her feelings to herself. His behavior was incomprehensible; men must be a different species if they can divorce their feelings from their sexuality with such apparent ease. She would have to ask Melanie about this later.

It was finally decided that they postpone their departure for two weeks. A crowd of people, made of their parents and relatives, escorted them to the airport. All the women were dressed in black and the men were wearing black suits and ties. For many months after his mother's death, Sammy would wake up in the middle of the night in a sweat, crying out.

"She's coming towards me in her shroud. She's pulling at me to take me with her." He wailed like a child. Mothers and sons—what fierce attachments they make. Nadia promised herself not to oppress her son with the burden of her love. She would try to let go of her children and let them grow without clinging to them like a deadly vine.

22

BEIRUT

A S HE CAME DOWN the airplane steps, he had been thinking, this is the beginning of my new life. He could hardly breathe and their bewildered child started crying because he was tired and restless. Sammy looked at Nadia, still wearing the black of mourning and looking miserable. She smiled back at him reassuringly.

The city of Beirut, bound by the lush mountains of Lebanon, was sweating in a Turkish bath of entrapped Mediterranean moisture. It was a sweltering September morning that greeted him with a wet and hot slap in the face.

"I thought that Lebanon was a summer resort?" he said. Sammy seemed to take the damp heat as a personal affront.

"Beirut is unbearably hot and damp in the summer, but the resorts high up in the mountains are always really cool," said Nadia.

Sammy was irritated by Nadia's response. Just because she had already spent several summers in Lebanon she was acting like a knowit-all. He gave her an angry look and then relented and forced a smile.

She was so brave to follow him on his new adventure; he must remind himself to always be kind to her. It scared him how much she trusted him and depended on him, and he secretly wished she were a little more independent.

As they drove into the city in a shiny new taxi that contrasted sharply with the rickety cabs that rattled around Cairo, Sammy became increasingly irritated.

"Switzerland of the Middle East is pretty shabby," he whispered uneasily, looking suspiciously at the sullen driver. Sammy's attention was focused on the spectacle of the Palestinian refugee camps situated on the outskirts of Beirut that blighted the horizon like a festering sore in a sick old body. The camp shanties and tents were a disgrace.

"Look at that. The Arab countries are too busy arguing among themselves and are waiting for the United Nations to solve their problems; look at this nightmare!" he said.

"They'll wait forever; nobody cares. It's not a very pretty sight," Nadia murmured in English. The driver, a fierce looking young man, sun tanned and muscular, was sweating profusely and cursing the other drivers as he weaved his way between the cars. He stank like a rotten fish.

"Where you from?" he addressed Sammy with a yellow smile.

"Egypt." Answered Sammy, disinclined to get into a conversation with him.

"*Ahleen, Ahleen*, welcome to Lebanon. Nasser is our hero. He's made us all proud to be Arabs. How is he?"

"He's just fine. Working hard on uniting all the Arab countries."

Sammy couldn't hide the irony in his voice.

"May God make him victorious against all our enemies."

The driver's black eyes darted at them in the rear view mirror, as he continued to furiously plow his way among the cars. There was a shiny brass tablet with an inscription from the Koran dangling from the mirror. Sammy found it a mockery that the shiny tablet proclaimed the man's creed in a country torn by religious factions and a civil war that had ended only two years before.

The drive was rough, just like being in a bumper car race, for there were no apparent laws that governed the drivers; they seemed to be demented by the heat and the heavy traffic.

"What have I done?" Sammy reflected, "dragging my family into this unruly land? Why, only a few years ago these people were at each other's throats; Moslems against Christians and Shiites against Sunnis . . .

I must be crazy; heaven help us in this lawless country. Switzerland of the Middle East, indeed! A few banks don't make a country. Here I am, a rootless immigrant, trying to start a life in this bad excuse for a nation."

Sammy was relieved when they finally arrived in front of their hotel.

He was actually surprised that they had arrived at all. In addition to all the frustrations, Ahab's whining had almost driven him crazy. The driver helped them with the luggage.

"How much do I owe you?" asked Sammy.

"It's on me. I'm happy to serve a brother from Egypt."

"But that's not possible," said Sammy, taken aback and feeling embarrassed by the unexpected generosity of the driver.

"It's for Nasser's sake. It's the least I can do for the people of Egypt." Sammy looked at Nadia and smiled.

"Thank you, my friend, what a nice welcoming gesture to a stranger."

"You're not a stranger, you're a brother. Are you Christian or Moslem?"

"I'm a Moslem," said Sammy after a slight hesitation.

"Are you Sunni or Shi'a?" The persistent driver asked.

"Sunni, most Egyptians are Sunnis."

"*Al hamdu lillah*! Thank God. Here is my card; if you need anything just call me. I'm always at your service."

"*Shukran*, Thank you."

"May your eyes be blessed."

That was the first quaint Lebanese expression they had heard.

Sammy smiled and relaxed a little.

The small hotel was right across the street from the beach. The three-story building was painted white and had the in progress look of a project hurriedly put together. It told the tale either of money running out before the job was properly completed, or of some other dire urgency that forced the owners to open up for business before they were ready. The outside entrance was bleak with no attempt at landscaping whatsoever. A young man in shirtsleeves, jeans and sandaled feet, picked up two pieces of their heavy luggage and escorted them into the hotel.

"What a lovely beach!" exclaimed Sammy to the young porter, who launched into a monologue delivered at high speed.

"St. Balash is the only beach that is free to the public. All the other Beirut beaches are private and have Saint's names, St. Pierre, St. Michel.

called Saint Balash, which means Saint Gratis. You don't have to pay."
Sammy laughed heartily but he saw that Nadia was too tired to be amused.

They entered the lobby which was small with a cheap-looking reception desk on one side. The wall was dark blue. To the left of the reception was a very large room that served as sitting room, dining room and bar. The walls were painted a powder blue, and though the smell of new paint was in the air, it was already peeling off the edges. The humidity must be merciless. The furniture promised not to last until the next tourist season. A few posters of Lebanese resorts decorated the walls and a huge garish painting of the Cedars of Lebanon hung behind the bar. In the dining area the chairs were covered in flowery pink and blue cotton and the tables were decorated with blue vases filled with pink artificial flowers. They registered quickly and were lead to the elevators.

They took the small elevator to the room that their sponsor, Mansour, had reserved on the third floor. The smell of disinfectant mixed with the smell of fresh paint was strong in the corridor that was otherwise clean and bare. There were no carpets on the tiled floors and this made the room look naked, but cool.

"It's only for a few days," said Sammy, almost to himself, as he swung one of the suitcases on the bed and went to open a window.

"Ahab is going to enjoy the beach. It's fine," said Nadia. She smiled at him and put her hand on his arm but she looked pale. His heart gave a flip as he looked into her pretty eyes.

"Don't worry, we'll find a nice apartment and I'll soon start working. We'll be okay, and if things don't work out, we can always return to Egypt." Sammy was only trying to reassure himself; his father in-law had called him a reckless adventurer; was he right?

Things seemed to go very well for them from the start. The producer, Mansour, was eager to help them settle down and begin work, and soon Sammy found a great apartment on the ninth and top floor of a seedy building in the elegant neighborhood of Hamra. The two-bedroom apartment featured an enormous terrace and a lovely living room. They quickly bought a small box spring and mattress for the boy and a kingsize bed for themselves. They also bought a table and four chairs for the terrace, which they used temporarily as dining room furniture. They would camp out and slowly furnish the place to their taste.

That first night Sammy woke up screaming and frightened Nadia out of her sleep. He had dreamed that his mother was coming towards him with her arms outstretched. His heart was beating fast and he had goose bumps on his arms. He was trembling.

"She's trying to take me with her."

"Calm down Sammy, you're just tired and nervous."

"Do you believe in ghosts?"

"No, I don't believe in ghosts, but it's normal that she haunts you. It's been such an unexpected shock."

"I'm going to take a shower to calm down." He got up and walked unsteadily to the bathroom. "Damn it!" he cried out. "There's no water in the pipes." He came back to bed naked and furious; he put his head between her breasts. "I left the tap open so we can hear the water when it starts flowing."

It was four in the morning when they heard the water running and Sammy got up to take his cold shower. It would become a habit; keeping the taps open all day and waiting for the water to arrive. Nadia learned to fill the tub every night for emergencies. They got used to flushing the toilet with a bucket of water.

Not only was there a shortage of water, the electricity was also temperamental and unreliable. It meant daily shopping for perishable groceries and sometimes walking up the nine flights of stairs because the elevator was stuck somewhere in limbo. They were young and didn't mind those inconveniences.

The elegant neighborhood turned out to be the red light district of Beirut. The narrow curving streets sloping downhill to the Mediterranean were full of nightclubs and bars that regurgitated an uninterrupted stream of drunken bodies all through the night until the early hours of the morning. From dusk to sunrise the noise was constant, punctuated by the shrill laughter of the daughters of the night and the grunts of their customers.

The building was a beehive of tiny apartments. Guessing from the activity going on, the rooms seemed to be rented by the week or maybe by the hour to a flood of young Arab men from all over the Middle East. Sammy later learned that most of the prostitutes in Beirut came from Egypt.

These women were not the most beautiful in the Levant, but were notorious for their sensuality.

It was a new life and a new start. Sammy was to begin immediately on a new movie. He was introduced to a Lebanese writer who arrived every day punctually at eight in the morning for breakfast, and then the two of them would sit on the terrace, at the round glass table, and spend the whole day working together. The script gradually took the shape of an action movie built around two popular Lebanese singers who had been contracted in advance. It wasn't an easy job. The stars were a heavy challenge, each of them weighing over three hundred pounds.

The two writers were undaunted by their enormous problems and soon managed to invent a love story full of struggles and feats of valor.

Nadia, who ran back and forth all day serving coffee and sandwiches, found herself sitting with Sammy and Shalaby, the Lebanese writer, listening to their discussions and even suggesting her own ideas. Sammy couldn't imagine seeing the two leviathans kissing or embracing on the screen so he refused to let the two stars meet. Every ten minutes, he managed to introduce a timely disaster that would quickly put a distance between them. That wasn't his only problem since the stars were popular singers; every scene had to have a song inserted at the crucial moment. It was not a musical, but an action film sprinkled with many songs and it was no surprise that, even though the movie had a happy ending, it never was a great success.

Sammy needed the money and couldn't refuse any job that was offered to him. He was lucky that his next movie was a great hit, and he soon became very much in demand.

CRICKETS

NADIA HAD, AT FIRST, found her new life very exciting. For the first time, she was away from the suffocating embrace of family and in-laws. She had hopes of finally finding herself. She could look at Beirut with fresh eyes not veiled by habit and taboos that governed every step of her life in Cairo. Lebanon seemed like a new world full of promise and adventure.

Sammy was doing the thing he loved best and Ahab was happy in school. She would take some time to settle down and then continue her studies. Most important, she was with Sammy and they would face the world together as a couple.

It was very thrilling to be invited, almost daily, to a dinner party at one of the well-known restaurants by the sea. They left Ahab with a babysitter and went out in search of what she thought was adventure.

Like an old whore, Beirut looked better at night when the wrinkles caused by unfinished constructions and the scars inflicted by piles of uncollected garbage disappeared like magic from the face of the city.

Soon, Nadia found herself sucked into a strange nightly routine.

Dinner never started before ten in the evening and always required formal attire. It was thrilling and glamorous at first, but after a few months of going out every night the ritual became no longer exhilarating, but a boring chore. At these dinners, Sammy always ignored her presence and was consumed by his business discussions. He insisted that it was an important part of his work. Nadia felt especially humiliated when he flirted with the pretty young starlets. He became more aggressive towards the girls, grabbing them and putting them on his lap. The company laughed and looked at her to see how she would react. She was not amused.

Nadia started making excuses for not joining him. He protested a little but continued his routine undisturbed. He always came home exhausted smelling of cheap perfume and whisky. He rarely made love to her anymore. He became like a stranger, totally consumed by his work.

One early evening as she sat on the terrace with her son, Nadia was lonely and sad. The green trellis that Sammy had put up all around the area and had seemed to enfold them in a protective embrace, now felt like a prison. A song on a radio across the street traveled, loud and clear, to the secluded balcony:

> *I carve your name on ancient oaks,*
> *You draw my name on shifting sands,*
> *Your name will stay for all to see*
> *My name will fade from memory.*

She used to love those romantic songs and her heart would break over the sadness of the laments, but now there will be no more dreaming for her. She felt caged in. Her career had melted away and her love for Sammy was dying slowly, but surely.

She lifted her eyes up to the sky and noticed that there were no birds of any sort to be found. The Lebanese, passionate hunters, had made sure that there were no birds left in the skies of Beirut. Nadia saw them, the hunters, every Sunday puffed up in their new hunting suits and carrying glistening guns. When they failed to shoot any birds, they stopped at the edge of the city to buy bunches of dead finches from young street boys. They hung their delicate prey from the radio antennas of their Jaguars and Porches as they drove proudly home to their families, bearing their bounty.

There were no more chirping birds, only thousands of screeching crickets in the pine trees, competing with the harsh sound of the waves breaking on the nearby rocky shore to create a relentless cacophony.

Her only consolation was that Melanie and Hassan, who were on their way back to Cairo, had accepted her invitation to come and pass a week's vacation in the quiet cool mountains of Lebanon. Melanie had said in her last letter that she was not feeling well and hoped that the fresh air of the mountains might do her good.

Ahab was busy making a huge fish out of *papier-mâché*. It was a school project that Nadia found herself fascinated with. They had both made a frame out of chicken wire and now she helped him cut strips of newspapers. They covered the paper strips with glue, as the teacher had instructed, and stuck them over the wire. It was going to be a clumsy, but beautiful fish. Nadia was contented when she had these peaceful moments with her son. These were perfect moments, when she felt that the world made sense.

Lebanon was artificial and harsh. A tiny land steeped in religious schisms and political corruption; it was a dangerous place to be, not a place to raise children. She could almost smell the danger. A place that had no birds was not a safe place. She had this vague plan to escape with Ahab somewhere, and the two of them would make a life of their own. Tomorrow she would go to the interview for the job at the library.

It would be the first step towards independence.

The interview at the library was a disaster. She couldn't type fast enough and didn't have experience in filing. Her bachelor degree was of no use for the job. The director offered her one hundred liras a month.

It would not cover the cost of a babysitter. She turned down the job and walked out feeling crushed.

The men ogled her on the street and made lewd remarks. A rich Arab in his shiny new Mercedes stopped his car and offered her a ride.

"How much?" he asked smirking.

The angry look she gave him only amused him and he persisted in following her. She decided that walking was not a good idea, and wondered if the only career open to her might be prostitution. She escaped from him into a hairdresser's salon.

Sammy was home when she returned. "Where have you been?" he asked angrily.

"I was at the hairdresser."

"Hairdresser? I don't believe you," said Sammy. "Are you screwing around behind my back?"

"If I want to screw around, it would not be behind your back." She was seething with anger.

"I don't trust women. They are all whores," he said.

"Sammy, I'm not a whore. I'm your wife and the mother of your son. You insult me and degrade me. I can't take this abuse any more."

"Darling, I'm joking. I didn't really mean what I said. Can't you take a joke? You know I love you very much. I'm just stressed out. Sorry my love, and if I hurt your feelings."

Later that evening Sammy asked, "Are you ready?" He was dressed up and ready to go out to dinner.

"I'm not going out tonight," said Nadia.

"Come on now. I said I was sorry."

Was this coarse brute the man she loved? How could he pretend that everything was okay?

"Mummy, a story." Ahab called from the other bedroom.

"Yes sweetheart, in a moment," Nadia answered.

"I have to go. It's part of my work. Mansour the producer and the stars of the film will all be there."

"I'd rather stay home tonight."

"Suit yourself. You don't really have to come," Sammy shrugged.

"I'll try not to stay out too late."

She knew very well that those dinners always ended up with breakfast at Marouche's, but she said nothing.

"By the way," he asked, "When are your friends coming? I have to reserve a room for them in Deir el Qamar."

"They're arriving on Friday; in four days."

"Good. I'm leaving after tomorrow and will send you a car Saturday morning to bring you all up the mountain. It will be fun for everyone.

The town was the residence of the governors of Lebanon in the seventeenth century. It is full of palaces and historical buildings, as Hassan will undoubtedly point out. The place is picturesque and the hotel is charming.

You can watch the shoot and Ahab can ride horses; he'd love the mountain."

It would be exciting to watch Sammy directing an action movie.

There would be hundreds of extras all dressed in ancient costumes, and horses covered in armor and draped in colorful banners.

They were living in two different worlds. Sammy was living his dream, but Nadia was not living at all. She was just a spectator. She rarely saw him anymore. He was always working and when he came home he collapsed and slept all the time.

That night Nadia waited for Sammy. She couldn't sleep. When he opened the door, dawn was breaking. He smiled when he saw her, then ran to the bathroom. Now she wasn't going to be able to talk to him.

She could hear him throwing up. Sammy could never hold his liquor but he always tried to keep up with the boys any way.

"We all went to *Marrouch* for breakfast," he said.

She smiled and forced herself to speak, "I worry about you. You are working too hard."

"If you want to worry, worry about the Egyptian secret service.

They came to the film set today and asked me to go to the Egyptian Embassy for questioning."

"Egyptian Secret service in Lebanon? What do they want from you?"

"I've heard that they ship troublesome people back to Egypt in wooden coffins," laughed Sammy.

"Can they do that? Lebanon is not part of the United Arab Republic. It's not under Egyptian control."

"There's no longer a United Arab Republic but the long arm of Nasser is everywhere," said Sammy. "We're naïve to think we got away.

We're like fish in a glass bowl."

"Oh my God Sammy, what did you do to make them treat you like a criminal?"

"They just want to pester me a little. They can't forgive me for leaving the army and more important, that I'm starting a movie industry in Lebanon. Don't worry, I'm taking two tough fellows with me. I'll call you as soon as I leave the Embassy." He laughed, "No one is packing me back to Cairo."

Nadia started trembling; she was scared and tired.

"Let's get some sleep, it's almost morning. Please don't worry baby, I'm not that important. They will not hurt me."

She still slept in his arms but she didn't feel safe anymore. Waking up later with heavy limbs was like waking under water. She struggled to move but could hardly budge. Nadia was paralyzed with fear; she couldn't face the day. She covered her head with the blankets and wished the day away, but soon realized that she couldn't afford to be depressed.

She would have to get up and take Ahab to school.

24

DEIR EL QAMAR

MELANIE AND HASSAN ARRIVED at last. Nadia was shocked to see the state her friend was in. Melanie had lost so much weight and she looked very ill. Sammy was not there, and Hassan was praying in the bedroom, so the girls had a quiet moment together on the terrace.

"What a lovely apartment! It is wonderful to be so high up and look down on the world below," said Melanie.

"You always see the positive side of things Melanie, there is much to be desired in this apartment and this neighborhood."

"Oh come on Nadia, you should be happy, and most of all because you have such a lovely boy. You are so lucky."

"I guess so, I'm not complaining," answered Nadia blushing with guilt. Hassan joined them soon after.

"Can't wait to see Deir El Qamar. I heard it was one of the most well preserved medieval cities in Lebanon," said Hassan. "I am especially excited to visit the Fakhreddine Mosque. It was built in the fifteenth century!" He has not changed much. Sammy was right, he was as pedantic as ever, thought Nadia.

The car arrived for them at ten in the morning and the group drove up the mountain to their destination. The town was certainly picturesque and the views breathtaking. The air was pure and smelled of pine. Nadia hoped it was going to be a great week for all of them in more ways than one.

The dinner that evening was in the huge sprawling restaurant that occupied the terrace of the hotel. The whole film crew was invited.

The pungent smell of *Arak* wafted through the air and mingled with the heady perfume of the pine trees. Nadia loved the smell; it was invigorating and sensual. The transparent liquor in her glass turned milky white when the waiter added ice cubes and water to it. The drink tasted and smelled of licorice. She sipped it slowly and decided that *Arak* was definitely an acquired taste. Every now and then Nadia let Melanie, who was sitting next to her, take a quick sip from her glass without Hassan noticing.

The star, the beautiful singer, Nagwa Amir was seated on the left of their host, Mansour Ibrahim, the film producer and Sammy's boss at the moment. He also happened to be the most handsome man Nadia had ever seen. Next to Nagwa sat Sammy, giving her his full attention.

To the right of Mansour sat the Turkish dancer Yeldiz.

"You are shining tonight," Mansour said to Nagwa, turning his back on Yeldiz and peering down Nagwa's milky white cleavage.

"You're the queen of the stars and a jewel among the ladies," Sammy in his turn said to Nagwa. It was obvious that he was being facetious, but Nagwa seemed to glow with happiness at the attention she was getting. Nadia blushed; the corny dialogue was embarrassing.

"Don't worry, these men are just being foolish. They always heap compliments on me but they don't mean a word they say," said Nagwa to Nadia, laughing. Nadia caught Sammy's eye, but he gave her a glance that clearly said "you're on your own."

"You are indeed a shining star," said Nadia to the lovely movie star.

"Good girl," whispered Melanie to Nadia.

Mansour turned to Nadia with his rakish smile. "How do you like your new life in Beirut, Nadia?"

"I like it very much, thank you." she answered, her heart beating faster than necessary. She was blushing and feeling like a tongue-tied idiot. He continued to smile at her. Nadia couldn't think of anything else to say and as no help came from Melanie, the conversation ended there. Nagwa's elegant chocolate colored evening gown contrasted dramatically with her flawless ivory skin and offered a perfect background for a display of an unusual set of topaz jewelry. The bleached-blonde hair, a coveted color in a country where most people had brown skin and brown hair, glowed under the lights like pure gold. She had a distinctive laugh, like the mating call of a hyena! It was a laugh that was designed to attract attention. Her perfume reached Nadia across the table; it mixed with the smells of food and tobacco to create a heady combination of scents.

Nagwa was a legend in the Middle East. She was now in her forties but was still at the height of her glory. Her clear melodious voice had earned her the title '*the song bird of Lebanon*', but she was also just as famous for her numerous and short-lived love affairs, her expensive clothes, and her fabulous jewelry.

Nadia looked quickly down at her own dress. She was wearing a cocktail dress from her trousseau. It was six years old and completely out of fashion. She was grateful that she could still squeeze herself into it. She distracted herself by watching the waiters flitting back and forth, dressed in white shirts, black pants and black bow ties. They were continually placing small white china dishes filled with delicacies on the table. Like all the other nights, it was going to be a Roman-like orgy of eating and drinking. The extra long table would soon be covered with a variety of *Mezza* plates. Nadia had counted up to fifty different kinds of appetizers. The sight was amazing, a legendary banquet from the tales of "A Thousand and one Nights!"

Hassan, who was sitting across from Melanie, was concentrating on his food and drinking water. Beside Melanie sat a nervous man who was fidgeting in his seat. He looked very slimy and devious. He didn't bother to engage them in conversation. His mind seemed to be elsewhere.

"Do you miss being in the lime light?" Nadia asked Melanie softly.

"It seems such a long time ago," Melanie answered. "I don't even remember."

Nadia felt bad at having brought up the subject. Hassan was glaring at her but Melanie seemed not to mind. The pretty blonde, who sat opposite them at the table, said out loud: "I used to be a dancer in the cabarets."

There was a nervous laughter all around. The fidgety man sitting next to Melanie stood up and said in a low voice: "This is my wife Romy, she is from Vienna. Can she sit between you ladies? Maybe she could talk to you both in a lower voice."

"Sure." said Melanie, "Come on Romy, sit here with us."

When Romy came over and sat down between them, Melanie whispered to her, "I also was a dancer, and I loved it."

"Really? What happened?" said Romy, looking surprised.

"I got married," shrugged Melanie. She seemed relaxed and happy for the first time since she arrived.

"I'm married too, to Obeid. It was a big mistake." She said pointing to the nervous man who now sat across from her, "I'm trying to be a good wife. No more dancing . . . but maybe I can act. The director promised me a role in the movie. I'm working on it." She winked at Melanie. Nadia blushed.

Nagwa was now leaning over Sammy and whispering into his ear.

Nadia could see that he was flushed with excitement. He loved beautiful women and adored them when they were both beautiful and easy. He looked very handsome in his new light blue silk suit and Sulka tie.

She felt the familiar twinge of jealousy tormenting her. Would this be another affair? She wondered. Sammy seemed to fall in love with every actress he worked with. The affair usually ended with the movie, and Nadia had decided to pretend that she wasn't aware that anything was going on. She had read and reread De Beauvoire's, *The Second Sex*, which preached equality between the sexes and free love. But Nadia still could not get used to the idea of a philandering husband; she still found it humiliating. She would have felt better had she known that Simone had not practiced what she preached, and had suffered agonies from Sartre's philandering.

Farida, Nagwa's silent sister and constant companion, sat on Hassan's right. She had obviously gone to some trouble to be as invisible as possible, but she was nonetheless a very attractive woman. With little make-up, short straight brown hair, brown skin, and a simple black dress, she was a subdued darker version of her flamboyant sister. There was a cynical smile fixed on her face. Nadia suspected that Farida resented being the handmaiden of her famous sibling. Nadia caught her stealing glances at Sammy from time to time.

How was Sammy going to resist the temptation of so many sirens around him? She had better lose some weight quickly, thought Nadia.

The competition was prodigious.

Mansour had welcomed her very formally and barely directed another word towards her for the rest of evening. She knew that he wasn't being rude, but just being respectful. On the other hand, he was staring at Melanie most of the time. He looked like a movie star right out of a gangster movie. Melanie smiled at Hassan, who was looking at her from across the table. Was he jealous? Did he think that Mansour was trying to flirt with his wife? After a few minutes, Hassan got up and came towards them. He circled around the table and pulled Melanie out of her chair. "Come on dear it is late, you must be exhausted," he said in a loud voice, looking at Mansour menacingly.

Romy and Nadia were left together to make the best of the evening.

Romy was tall and slim and wore barely any make-up at all. She looked a little washed out next to the gaudy mid-eastern females with their elaborate hairdos, heavy makeup and round bodies. She was wearing grey pants and a white shirt and hardly any jewelry except for a thin gold chain around her neck. She was beautiful and radiated a subtle sexuality.

"Is this your first time in Lebanon?" Nadia asked.

"Yes, my husband Obeid is production manager in your husband's new film."

"Of course, I just realized that."

"We got married recently," said Romy.

"Congratulations," said Nadia. "How do you like living in Beirut?"

"Yes, not bad, but Athens was better."

"How long did you live in Athens?"

"Three years. I had a rich lover who really spoiled me. We had a lot of fun until one day he was arrested, and I found out that he was a gangster."

"Romy, what an adventure your life is!" exclaimed Nadia.

Romy shrugged her shoulders. "I spent all the money he gave me paying musicians to serenade him under his jail window."

"What happened when you had no more money?"

"I joined a group of dancers and travelled around the Middle East; performing in night clubs. I went to Cairo, Damascus, and then Beirut, where I met Obeid and married him. I thought he was a rich producer," she laughed. "I used to work as a stripper in nightclubs." Romy laughed and helped herself to some *kebba*. She ate with relish, the raw meat that was pounded with cracked wheat into a pink paste. It was a Lebanese steak tartare that Nadia was scared to eat.

"Wow, stripping must have been a hard job. You know, I've never even worn a bikini!" said Nadia, examining Romy with interest and blushing.

She tried to hide her embarrassment by gulping down some pressed caviar on buttered toast.

Romy shrugged, "Everything is hard at the beginning, but then it becomes easy, normal."

For a slim girl, Romy had a huge appetite. She was devouring now with gusto a boneless pigeon stuffed with ground meat, raisins and pine nuts.

The women's conversation was interrupted by the arrival of a tall and extremely handsome man who was accompanied by a very short and round person. Obeid immediately jumped up, and with a look of relief introduced the two men to the guests as Borham Isaac, the Turkish star, and Omar Arkan, his agent. The men made the round of the table shaking hands with the men and kissing the ladies' hands.

When Borham reached his compatriot, the dancer Yeldiz, she greeted him in Turkish and pulled him down into Hassan's empty seat. She gave him a big hug and a kiss on the mouth. It was obvious that they knew each other. Omar plunked his round body down into the chair next to Obeid and disappeared into oblivion. No one took any notice of him after that.

All the women and some of the men at the table were drooling over Borham. He was like a Greek God. He had curly blond hair and huge hazel eyes. Nagwa started giggling hysterically; she didn't bother to hide her admiration for the handsome actor. She whispered something to her sister to which the other one answered with a smile.

Nadia didn't quite catch what was said but she watched with interest as, in a very quick maneuver, the two women exchanged places and Borham found himself sandwiched between Nagwa and Yildiz. Sammy looked over to Nadia, smiled his most wicked smile, and shrugged his shoulders slightly. He seemed to be saying to her that even though Nagwa had abandoned him, he wasn't upset. It was just a game. He always likes to share his adventures with me, thought Nadia.

"Now watch the fireworks," Romy said to Nadia as she helped herself to some smoked salmon and a sip of *Arak*. "It's getting interesting, yes?"

Nadia couldn't hear the conversation across the table but it didn't matter, not many words were being exchanged. Nagwa spoke only Arabic and Borham spoke only Turkish, and yet a very subtle dialogue was going on. Nagwa, who was at least fifteen years older than the man, put a possessive hand on his thigh and completely monopolized his attention. Sammy had to content himself with talking to Yeldiz, who sat glaring silently at the absurd pantomime going on between Borham and Nagwa, barely responding to Sammy's valiant attempts to distract her.

Nadia and Romy focused on the delicious Lebanese dishes placed in front of them. They delighted in the chicken kebobs with white truffles and the stuffed grape leaves. The fragrance of mint in the salad blended with the smell of parsley and yogurt. Nadia wondered how on earth was she ever going to lose weight?

"Yeldiz tried to arrange a threesome," said Romy smiling. Shockingly, her lovely mouth revealed yellow teeth.

"With you and your husband?" Nadia asked, blushing deeply.

"No, silly, with me and Mansour."

"Be careful, Romy. These are dangerous games."

"Nadia, you're such a baby." Romy burst out laughing. "Yeldiz is his mistress at the moment. Don't you think he's a very handsome man? Mansour is the richest man in Beirut," murmured Romy. "He has many companies."

"Really!" said Nadia. She couldn't help being intrigued.

"He is also a great lover but I don't like Yildiz, and *en plus*," continued Romy, "I'm trying to be a good wife. It is so hard to be faithful to one man."

"I don't know," answered Nadia laughing, "I've never known another man except my husband."

Romy gave her a serious look. "Just look at all the great dishes around you! You want to tell me that you like to eat steak every day and are not tempted to taste all of these other good things?"

"Men are not food. There are feelings involved; love, commitment, friendship, fidelity . . . , said Nadia, laughing nervously. "Besides if you eat from every dish you might get a stomachache."

"Food poisoning, I think," said Romy laughing.

"Exactly," agreed Nadia.

"Right now I'm on a diet," affirmed Romy. Nadia felt like a fish out of water. This conversation was too disturbing. She excused herself and went to the ladies room. There, she had to wait for a booth while a couple of women were busy vomiting the food they had gorged on during the evening. So this was their secret diet.

When Nadia returned to the table she felt like a stranger crashing the party. Sammy was too engrossed with Yildiz and didn't even look at her. She wanted desperately to leave, but was afraid to disrupt the evening and anger Sammy. It wasn't good for business. She sat down and asked for a fresh glass of *Arak*.

All those people seemed to Nadia like characters in a novel. They fascinated her with their exotic plumage, unconventional behavior and uninhibited language. It was interesting to watch them closely. Like reading a book by *Zola* or a novel by *Hemingway*. Maybe she should start writing a book about the crazy life in Beirut, the Switzerland of the Middle East.

Yeldiz smoked a *narguila*, making a bubbly noise as she drew the smoke through the water to her lips. She sat there smoking her water pipe and eating a barbecued lamb chop with her bare hands; the grease running down her chin. With her wide black eyes, she stared at the food like a wildcat. Her curly hair was shoulder length and untamed.

Her plump mouth was large and greedy. She wore a shiny dress with a deep-cut neckline. Her round breasts seemed about to pop out of her bodice. Nadia found herself searching Mansour's face for scratches from the wild cat.

Yeldiz was still trying desperately to engage Borham in a conversation, but was outmaneuvered by the aging singer. The duel between the two women was amusing. Nagwa suddenly stood up and lifted her glass of *Arak*, saluting all the diners in the restaurant. They all clapped and cheered her. Someone begged her to sing a song and she immediately complied. She launched into one of her most popular ballads *a capella* and her strong melodic voice rang through the restaurant like a silver bell. The diners went crazy with pleasure, clapping and asking for more.

When Nagwa finally sat down flushed with excitement, Yeldiz jumped on the table and started dancing among the half empty dishes of *mezza*. She danced to the rhythm of the clapping and drumming on the tables by the guests. By the time Yeldiz sat down sweating and exhausted, the sight of her rival kissing the beautiful Borham made it clear who had won the battle. Mansour seemed very amused by the spectacle.

Sammy was getting slightly drunk. Nagwa had dumped him for the Turkish actor. He was peeved but his vanity prevented him from showing anger, and he concentrated on convincing Mansour to start funding for the next movie.

Nadia was looking at him imploringly. He blew her an encouraging kiss then whispered to Mansour, "I'm sorry, but I would like to be excused, I have to get up early tomorrow."

"Of course, I understand. May God be with you brother."

Sammy got up and excused himself to the company. Nadia joined him after saying swift goodbyes all around and they hurried away from the dinner party.

The scene they were shooting the next day was in a cave as big as a city square. The gas flames lighting the damp walls projected shadows on the twenty or so horses and their riders who were standing at one end, waiting for a signal. Sammy called out "action!" through his megaphone, and all horses galloped across the camera and out of the dark cave into the daylight. The scene was repeated over and over again, until Nadia felt that the moisture from the cave walls penetrated into her bones. She decided to get out of the cave, but Ahab refused to leave his father. The boy already wanted to be a director and was holding his father's megaphone with both hands. Nadia ran out of the dark cave as if pursued by demons. Her nerves were shattered since the incident with the Embassy. The interview did not turn into a kidnapping this time, but you never knew with these people. She was still worried about Sammy's safety.

The sun greeted her with a warm embrace, and the sweet smell of the pine trees was refreshing. Melanie and Hassan were sitting outside in the shade of a pine tree. They looked like sad lovers who were meeting for the last time. Hassan never left Melanie's side for a second. He touched her all the time, as if he wanted to make sure that she was real.

Melanie had lost all her lively spirit; she was like a puppet hanging from her master's strings. Hassan called Nadia over.

"We have been waiting for you. We have something to tell you, a surprise." There was a dramatic pause. "Melanie has changed her name; she is now Fatma al Zahraa, like the daughter of the Prophet,"

said Hassan.

"What? Why change her name? Melanie is lovely." Nadia looked at her friend, bewildered. She remembered how Hassan didn't even shake hands with her when she met them at the airport. He was becoming such a fanatic Moslem, afraid of getting polluted by a dirty female, she supposed. Changing her name? That was a bit too much.

"Remember when she was once Soraya Salem, the folk dancer?"

Hassan laughed. "She's used to changing her name to fit the occasion."

"And what was the occasion this time?" said Nadia angrily.

"She is married to a good Moslem now. She is now my wife." Nadia blushed in anger. She would have liked to break his nose. There was no need for him to be so mean. She was glad to notice that he was starting to lose his hair, but regretfully, none of his animal attractiveness.

There was always a mean streak in that fierce tiger. She looked at her friend, expecting her to lash back at Hassan and defend herself, as she had known her to do, but Melanie was pretending to not hear their conversation.

Hassan finally got up and walked away towards the hotel, leaving them alone. He was always such a prick. People don't change, do they?

People are supposed to mature, to improve. "No, they don't change," Nadia thought in despair. She continued to sit with Melanie under the pine tree. Something was seriously wrong. Nadia had been pretending all along not to notice the ribs that were sticking out of her emaciated friend's flat chest. Melanie looked very ill.

"I must look terrible," said Melanie. It was as if she was reading Nadia's mind. "Please wipe off that frightened look off your face. I'm really okay. I've been very ill, I had pneumonia, but I'm getting better. Don't worry."

"Why didn't you tell me?" Nadia started to question but stopped herself. "The mountain air will do you good and the food here is so great. We'll put some flesh on these bones yet," she said, trying to hide her anxiety.

"How time has wings. It seems like yesterday when we left for Moscow. You were about to give birth to Ahab and now he's almost seven. I can't believe it's 1961 already. Pretty soon we'll all be old; how sad. Unless, of course, we die young."

"Why are you being so morbid? Is this the influence of communism or of the Russian cold winters?" Nadia tried to be light hearted, but a sense of foreboding crept into her heart.

"Lunch break!" Sammy's loud voice was heard coming from the cave. Nadia looked towards the cave entrance and saw Sammy coming out with Romy holding on to his arm. He had given Romy a small role in the movie. Nadia's heart leapt into her throat. Was he going to have an affair with Romy? She really looked stunning dressed as a Bedouin with all the makeup and jewelry they had put on her. Was she the new candidate? Then a parade of the cast and crew began to come out of the cave following Sammy.

Mansour was riding in his red Ferrari with Yeldiz sitting next to him. He stopped and Romy jumped in. Here was the *ménage a trois* that Nadia had fantasized about. The threesome waved goodbye, as they roared up the hill to the hotel. Sammy seemed unmoved.

Obeid came out of the cave, wearing his dark eye glasses and smoking a cigarette. He was talking to the stout makeup woman.

They walked slowly and behind them was a young assistant carrying a heavy makeup case.

Several horses came out being led by their trainers. Borham came cantering out on a black steed. A camera man sauntered past carrying some gear. His assistants pushed a trolley with a heavy camera on a tripod. The extras were laughing and jostling each other.

Finally, Ahab came out from the cave behind them, riding a white horse with one of the gaffers holding the lead. "Look Mummy, I'm a chief," he was wearing an actor's helmet.

"You look really fierce on that big horse."

They all slowly walked the two hundred yards uphill towards the hotel. The narrow road curled up between the mountain on one side and a steep valley on the other. It was lined with old pine trees that filtered and perfumed the air. The breeze was cool with the sun barely sneaking through the brilliant green branches of the trees. The crickets continued their endless song, oblivious to the people and the horses walking on the road.

Nadia walked with Melanie slightly behind the group.

"What are your plans for the future?" she asked Melanie in a whisper.

"We're going to Cairo for three years until they appoint him somewhere else."

"What about you? Your own plans?" Nadia insisted.

"I'm planning to get well and then we'll see. I'm not important."

"You always advised me to be independent. What happened to you?"

"I don't know. I love him Nadia and I'll stay with him until the end."

"Melanie, that's not what I expected from you. Life is not what I thought it would be. I'm miserable." She stood there with her arms around her friend, while scenes of the past seven years unrolled like a bad movie in her head.

"No sweetheart, life isn't exactly what we read in Jane Austen is it?"

said Melanie. "Life is a novel written by a careless author. There's no outline and no logic to the events."

"I'm afraid you're right. I don't see any happy endings anywhere."

The two women continued walking and soon the rest of the group entered the hotel.

"Nadia you should be very happy. I keep on thinking what a lovely apartment you have in Beirut. You don't realize how many people are living in dark rooms, yearning for the light of day. Besides, you have a lovely boy and a loving husband. You have a lot to be grateful for."

"That was all I ever wanted out of life, but now I feel cheated. I can't depend on my husband's love. I want to exist on my own merit and not at the mercy of someone else. Is that very evil of me?"

"We make choices when we're too young and unprepared, then we spend the rest of our lives regretting the path we chose. You have a good life. Try to make the best of what you have."

"And you, do you have a good life?" asked Nadia.

"The prophet said that the best life for a woman is to serve her man."

"The prophet? Are you kidding me? Melanie, are you joking?"

"I guess I've changed. Hassan is my life now."

25

LYCHMANIA

"It's called Lychmania," he said; his face contracted with pain as he rubbed his unshaven beard. "It's caused by the bite of a horse fly."

Sammy smiled at her, that smile that had made her fall in love with him many years ago.

Nadia sank down on the sagging cot that was placed beside his hospital bed. She guessed it was placed there for her to sleep on. Sammy was lying on a white metal bed, dressed in a white hospital gown, and was partially covered with a white sheet. The brown military camelhair blanket, lying on top of everything, was the only touch of color in the room and echoed the burnished copper of his suntanned skin.

"Is it serious?" she asked, as she looked around the small room. She noticed a crucifix above the bed on top of a plaster sconce with a bulb sticking out of it. In order to avoid looking at Sammy's face, she fixed her gaze on a picture of the red ruins of Petra that was hanging on the wall. She realized that she was biting her nails and forced herself to stop.

"I almost lost my leg from the infection; it nearly reached the bone," he said. "I was fortunate the doctors here were familiar with the disease."

"I guess you're lucky." Nadia felt a wave of nausea. Her hope that she wasn't pregnant was fading away rapidly. He didn't notice her malaise; he was fidgeting nervously with the sheets.

"Lucky we finished shooting when we did. As soon as I'm well I'll have to start editing," he sighed. He seemed tired and distracted, but still looked handsome in his white hospital gown.

"I have a feeling it's going to be a flop," he said, laughing. He had a boyish laugh that disarmed anyone who came into contact with him.

When did they stop loving each other? She wondered.

"Why do you say that?" she asked.

"Oh, just a feeling. The chemistry just wasn't there," he laughed again.

A few days ago she had read in the scandal column of the *Beirut Times* that Sammy was having an affair with Romy. No names were mentioned, *Egyptian director seen constantly in the company of Viennese starlet.* The news wasn't a surprise, but reading about it in the paper stung her with shame. She remembered what Romy had once told her that love was just a game. She was tired of his games, but when he called her from the hospital and asked her to come over, she had, nevertheless, left Ahab with a babysitter and flew from Beirut to Amman to be with him. She hadn't seen him for several weeks.

"What does the doctor say?" She had to keep her voice steady and the conversation going somehow, or she would break down and cry.

She had survived the worse moments of despair. Once she had even stood on the balcony looking down on the street, nine floors below, imagining what it would be like to lie splattered like a cracked egg on the pavement.

"I'll survive," he said flashing his crooked smile.

"Yes, I'll also survive," she thought. She had been spending her evenings reading; she had begun to actually enjoy the solitude. She had started gathering information for college scholarships in the U.S. Melanie had suggested a couple of schools and had promised to help her get the admission papers. Making plans for the future had given her a sense of direction. At least in America she would be a whole person, not the half person she was in the Arab world.

"It seems to be a good hospital," she said. It was a Catholic hospital run by some holy order. The hospital room was narrow, small, and stark with a window overlooking a dusty road. There was a knock on the door, and a nun in a white habit, came in carrying a surgical tray.

"It's time for your shot," she mumbled, holding her head down.

Nadia couldn't tell her age, her face was hidden.

"Sister Mary, this is my wife, Nadia."

She raised her head. "Nice to meet you, Nadia." She had a distinct British accent. She was young and pretty with a plump and dimpled face. She smiled at Sammy, as she waddled to the bed with her loaded tray. It was obvious that he had already charmed her. He had a way with women. He did it effortlessly.

"Don't worry dear, we'll get him on his feet very quickly," she told Nadia.

"Thank you, Sister," said Nadia, unsure of how to address the nun.

The woman proceeded to give Sammy his shot. He hated shots and was afraid of them. He winced when the needle went into his skin.

The nun swabbed the punctured spot. "What a lovely wife you have, Doctor!" she said.

"Yes, I am lucky, and you are also lovely, sister."

She blushed and floated out of the room shaking her head at him.

"You look great," he said, arranging himself on the bed and examining Nadia from head to foot. "Lost weight?"

"Yes, about ten pounds." He was obsessed with her weight. It always irritated her when he mentioned it. She felt her face slowly turning red with anger.

Nadia heard a car coming up the quiet road; it stopped at the gate of the hospital. She got up from the sagging cot and looked out the window. She saw the red Ferrari, it was him. In Beirut, he had called every day to ask about her and Ahab while Sammy was away, and she had declined his numerous invitations to dinner.

He was arriving to check on his sick director, or rather on his investment. From the hospital window she could watch him without being seen. The car door opened and a freshly shaven Mansour jumped out of it. She guessed that he would be staying at the luxurious Jerusalem Hotel about two miles away. Why did he show up? She had hoped to be alone with her husband for a serious talk. He was a very troubling presence in her life, and he was her husband's boss. It irritated her that she found him attractive, but most of all, it bothered her to be beholden to him for their livelihood.

Nadia kept silent during Mansour's short visit, leafing through a woman's magazine she had bought at the Beirut airport. The blood was pounding in her head.

"Nadia, please, feel free to call me if you need anything while Sammy is away in Rome," he told her.

"Rome?" Her head jerked up from the magazine.

"Oh, so he hasn't told you yet?" He laughed. "He's going to Italy to edit the film." Mansour was smiling a mischievous smile.

She was confused and looked at Sammy who was scratching his beard, looking annoyed.

"It'll only be for a few weeks," Sammy said.

"That's a surprise," she suddenly felt sick again, but she managed a weak smile.

"Sammy is full of surprises," said Mansour laughing.

"Thank you, Mansour, for everything, we owe you a great deal," said Sammy.

"Just deliver me a good film, ok? I'll leave you two lovebirds together. See you in Beirut," and he left.

"You're quiet Nadia. What are you thinking of?" asked Sammy as he was fidgeting with the dials of the radio by his bedside.

"Thinking about what to do when you're away. Thinking about what we're going to do for money."

"Can you tell me how you could be so brainless?" he asked turning towards her. "It's not my fault that my wife was stupid enough to deposit my big paycheck in a bank that was folding. How could you? Just tell me, how could you?" his voice was getting louder.

"I didn't know." She sat there like a scolded child biting her nails.

"Didn't you read the papers about Intra bank? Didn't you see, while you were there, that the people were actually taking out their savings from the bank?" His face was turning red.

"I'm sorry. I had no idea."

"You're such a dim-witted child. When will you grow up? I'm tired of taking care of you all the time." He stopped yelling and said more quietly; "would you get me the pissing pot, it's in the bathroom. I need to piss."

As she approached him with the glass urinal in her hand, she smelled the putrid smell of puss and blood that was oozing from his leg. She felt another wave of nausea and tried to pull back but checked herself.

"How is Ahab?" he said, his urine making a tinkling noise against the glass.

"He misses you. It's been three weeks since he saw you."

"I miss him too."

A long silence ensued as Nadia took the urine to the bathroom and dumped it, flushed the toilet, rinsed the urinal, washed her hands, and returned to the room. Sammy's eyes were closed. She sat down again on the cot and stared at the blank, uninteresting white walls. Everything was white in that room. It was the Moslem color of mourning, the color of death, the color of shrouds. Sammy was lying down with his eyes closed like a corpse. She hated the color white.

"Are you having an affair with Romy?" she surprised herself by asking.

"What difference does it make? It's not important," he said.

"Are you in love with her?" she said.

"I think I am. I don't know." He said.

"Sammy, I see that you don't love me anymore!" She started crying.

"Oh God, I must stop being such a pitiable fool."

"Of course I love you, but that doesn't mean I can't enjoy other women. Life is too short, and I'm going to live it to the fullest."

"You make me feel so unwanted, so inadequate. What's wrong with me that you keep sleeping with other women?"

"Nothing is wrong with you. I love you, but I want to be free. I don't want to be tied down. You have so many sexual hang-ups, it's pathetic. You should take a lover; loosen up a bit."

"I don't play the role of the temptress very well. I can't sleep with the first man that crosses my path just to feel free."

"It would do you good."

"You're crazy." She thought of Mansour and his insistent courtship, but quickly dismissed him from her mind. "Neither do I like to play the role of the faithful wife, waiting for her wandering husband to come home." She decided to change her plans and to fly back to Beirut that same evening, to sleep in her home with her son. She kicked the cot with her foot in anger and started folding up the covers and the sheets that were on it.

"Would you rather I lied to you and had an affair behind your back?" he said.

"I'm old fashioned. I'd rather you were faithful to me." She slowly put the sheets on a chair and started to fold the cot. "It doesn't matter anymore," she said.

"What do you mean?"

She hated her weakness, yet her eyes filled up with tears.

"Sammy, I'm going to Cairo to see the family. It will give us time to decide about the future of our marriage."

"You're my wife and I'll never leave you. That much I can promise."

He stopped talking and lay back with his eyes closed. "I'm sorry, I never meant to hurt you."

"You sound like a stupid character in one of your awful movies."

She really wanted to hurt him now. She pushed the folded cot towards the door.

"Well, maybe it's a good idea for you to go to Cairo after all." There was no love in his eyes, only cold anger. "I just hope you don't get caught up in another war while you're there."

"That would be convenient for you." She put the bed linen on the cot and pushed it out of the room into the narrow hallway.

"Don't be stupid," he said. "You're being childish."

He turned his back to her and pretended to go to sleep. The heat was becoming oppressive. She felt a pain in her chest and a paralyzing numbness spreading all over her body. She became aware that what she was feeling was intense grief.

"By the way," she whispered with difficulty. "I'm pregnant."

"Damn it," he turned over quickly, wincing from the pain in his leg. "I can't shake hands with you without your getting pregnant." He pushed the dish of fruit that was perched on the side table onto the floor, and it crashed into a thousand pieces. The oranges rolled on the floor and under the bed. She didn't make an effort to pick anything up.

"Don't worry," he said after he calmed down. "I'll find a good doctor and take care of this as soon as I get out of the hospital."

"No," she said in a low voice, "I'm keeping this child."

26

A CALL FOR HELP

IN CAIRO, MELANIE WAS confined to Hassan's old bachelor room. It was old-fashioned with a four-post brass bed that jingled when they made love. She often thought she heard muffled footsteps in the corridor.

When she made the effort to tiptoe to the door and open it, there was no one there. At times, she was certain that she was being watched through the keyhole. She even once spied a face in the window behind the translucent curtains. But no, there was no one there either; it was only the breeze. Was she losing her mind?

She thought of confessing her fears to Nadia, but Nadia was in Beirut, so far away. Besides, Nadia thought of her as a pillar of strength and would never take her fears seriously. As for Hassan, the knight in white armor who was going to save the world by the force of his true religion and social justice; he would tolerate no weakness.

She sat in the large overstuffed armchair with a book in her lap, unable to read. The walls of the room seemed to lean over and close in on her like the walls of a jail. They were painted in an ivory color that had yellowed with time, and were decorated with Hassan's many awards and certificates all framed in black. She had spent many hours contemplating those awards and had memorized them, including the framed Koranic verses hand-written in gold on a black background.

Melanie yearned for some of her own wild colorful paintings to cheer up the room, especially the one that Nadia had painted for her of a bare-breasted woman carrying a basket of mangoes.

"She looks just like you," she had told Nadia.

"Of course, it's a glorified self portrait, silly," Nadia had answered laughing. "Let's put it up in your kitchen. We don't want to shock your visitors."

"To hell with my visitors," Melanie had answered. "I love it and will place it in the living room."

How her life had changed since then! Nude or covered up, it was against Hassan's faith to hang representational art, he absolutely forbade it. Nadia's painting had disappeared. The odors of the house invaded the room through the walls. It must be Tuesday; she could smell the soap water boiling for the laundry. The radio was on in the living room and she could hear Fairuz singing:

> *Give me the Nay and sing*
> *The secret song of eternity.*
> *The laments of the flute will linger*
> *Beyond the decline of existence.*

The words of Khalil Gibran, sung to the sad music of the flute, brought tears to her eyes. She remembered her days of glory when she was dancing on the stage every night, and her body was like an instrument that vibrated to the music. She had felt powerful then and had never thought of 'the decline of existence.' She stood up and twirled to the music, but lost her balance and fell back into her chair, seized by a coughing spell that left her exhausted.

"This will do you good, my dear. Finish it all. You have to regain your health," Wafeya said, as she came in with the chicken soup.

"When are we going to find a place of our own, darling?" Melanie asked Hassan later when he returned from work.

"What did you say?" he asked, never looking up from the paper he was reading.

"I need my own apartment; I need some privacy."

He glared at her. "The woman is serving you hand and foot and all you want is to be on your own." He returned to his paper, concentrating on whatever he was reading. She kept quiet for a long while counting the seconds ticking in the clock behind him. His stiff body was twisted to one side away from her. She noticed that his shoes were covered with dust and wondered if his mother was going to shine them for him.

When she looked up, she became aware that he was not reading at all; he was staring at her over his paper.

"I wonder sometimes about what goes on in your head. Don't you know that the Koran says:"

> *In travail upon travail*
> *did his mother bear him,*
> *Show gratitude to Me and to your parents.*

"I have to take care of my mother." His foot was tapping the floor slowly.

Melanie started to cough and couldn't seem to stop. She gulped down some cough medicine and slumped back further in the huge stuffed armed chair, pale and exhausted.

A look of despair appeared on his face. "I'll find a place for us as soon as you get stronger. Just be patient." He bent over and felt her forehead. She slowly lifted her face to him, but he only gave her a quick kiss, stood up, and searched in the pockets of his jacket. "I almost forgot," he pulled out a crumpled letter. "There's a letter to you, from Nadia." He tossed the opened letter into her lap. She wasn't angry that he had opened her private mail. She didn't care anymore.

"*Dear Melanie,*" Nadia wrote in her large, even handwriting. "*Thanks for the information from UCLA. I am so excited about the prospect of leaving and starting a new life. Sammy will soon be going to Italy to edit his movie. I'll be coming to Cairo with Ahab to get my papers in order. I yearn to see you. I miss the old times when we chatted for hours about everything and about nothing. I need to talk to you about some serious matters. I hope you are getting over the pneumonia and are stronger than ever. How are you adjusting to living in Cairo again? I wouldn't be surprised if you are already involved in some artistic project. I'm dying to hear the exciting details of your life. Prepare yourself; I have some surprises for you. See you soon. I love you. Nadia.*"

Melanie smiled. It would be wonderful sharing her thoughts with Nadia.

Hassan was pacing around the room like a caged animal. She looked at him, questioning him silently, following him with her eyes.

"Nadia is coming to Egypt for the summer," she said, as if he didn't already know. He hesitated for a moment, and then nodded his head in agreement. His hands were thrust into his pockets.

"You have been helping her in secret. I hope you know what you're doing."

"Hassan, she needed my help. Her marriage is not working out."

"How about our marriage? Is it working out?"

"I love you very much. I'm sorry I'm always sick."

"It is not your fault. By the way, I was promoted to a very important post," he said.

"Where are we going to be posted this time?" She sat up straight, animated at the prospect.

"No, my dear, it's not exactly a consular post. I'm going to be the personal assistant to the President."

"That sounds very important." She slumped back into the armchair.

"It's a great honor. You must be proud." She had never imagined that he was so close to the top brass in the revolutionary council.

"Yes, it's a very important job. It's almost like being a minister." He frowned. The crease on his brow was becoming a permanent feature.

"Aren't you pleased?" She, herself, was very disappointed. She had wanted to be alone again with him in some foreign land.

"Yes and no. It's a very big responsibility and I'm afraid that I might fail." His eyes were on fire. He was a beautiful animal, her fierce tiger, and she loved him, even if he was killing her.

"You will not fail; you're so dedicated to the cause."

"I would give my life willingly to the president, but I'm beginning to have doubts about the revolution."

"My poor baby, you must not lose faith. You must be strong."

He looked at her for a long time with those eyes that had stolen her heart and robbed her of her free will. He didn't reply. He reminded her of the revolutionary hero in Conrad's *Under Western Eyes*. He had the same single mindedness and total dedication to the ideals of the revolution. Was Hassan also capable of killing for the cause?

"I have work to do." He left the room abruptly and went to his office to read the endless reports piled upon his desk. She wanted him near her but was afraid to ask. He'll creep in beside her late at night and hold her in his arms. She just had to wait.

There was a fallen tree across her path. Melanie found herself in an overgrown garden full of weeds and dead branches. She was hiding from someone but couldn't remember who it was. She heard a noise like that of a stalking animal, and she cried out in fear and ran as fast as she could before she stopped to take a breath. Suddenly, the dwarf jumped at her from behind the tree and embraced her in a strong grip.

She screamed, but he held her tighter. He was shaking her and shaking her violently. She woke up shaking in Hassan's arms.

"It's just another nightmare. I'm here beside you dear. Please, try to sleep." She was sweating and Hassan caressed her damp head and face.

He gave her some water to drink. She couldn't sleep. She felt that she would never sleep again.

The next day she found her old telephone book under some of her still unpacked clothes and phoned her psychiatrist. She was surprised when he answered the phone.

"Melanie, how are you, my girl? It's been many years."

"It's been six years and three months Dr. Anwar," she laughed nervously. "What have you been doing all this time?"

"Remember you're the patient my dear. I'm the one who gets to ask the questions."

She was silent for a moment, overcome with emotion and not knowing what to say.

"Have you been taking your medication Melanie? Are you well?"

"I got married. I was very happy and then . . . we went to Russia and I caught pneumonia . . ." She knew she was being incoherent and stopped to think of what to say next. "I've been very ill. Things are not so good right now." She was breathing hard. "I need medication. I need your help."

"I want to see you right away," he seemed concerned.

"My husband doesn't know that I have any problems. He wouldn't understand."

"It's not a crime to be ill. It's just a chemical imbalance in the brain. Actually, it would be better if I could see both of you."

"Please, help me. I need to sleep. Can you just send me a prescription to the house? Please, Dr. Anwar, I'm desperate."

"I'll give you something for just a few days, but then you have to come over with your husband. I'll send the medication with a messenger right away."

"Oh, you are so kind. Thank you, thank you, thank you!"

"What's your address?"

She gave it to him. Melanie thought that she heard Hassan's mother shuffling behind the bedroom door, but she dismissed her suspicions this time as paranoia. When the messenger arrived, the medication was confiscated.

RETURN TO THE YELLOW CASTLE

A FEW DAYS AFTER HER arrival in Egypt, Nadia sat drinking Turkish coffee in the Cairo Hilton cafeteria and waiting for Melanie. The architect of the Nile Hilton used Ancient Egyptian art as an inspiration but ended up with a garish Hollywood setting of turquoise mosaic and golden plaster. Nadia hated it. Nadia noticed two teenage girls sitting at the table next to her. They were in deep conversation, laughing, interrupting each other. They were exchanging secrets and giggling like young girls do. They reminded her of Melanie and herself at their age when they would chat for hours about what they thought were important issues. She was trying to listen to their chatter when Hassan arrived alone.

"Where's Melanie?" she asked.

"She's not coming, she's very ill. Actually, she's in the hospital." He murmured.

"Which hospital? What happened?"

"She's at the Yellow Castle."

She put her hand on the table to steady herself. "The Yellow Castle? What do you mean?" she asked, not comprehending. The memory of the poor demented patients roaming like animals in the yard of the mental hospital came back to her with painful clarity after so many years. The high

203

school science trip to the Yellow Castle had traumatized both Melanie and herself for a long time.

Hassan was talking, telling her that Melanie had had a nervous breakdown and had to be confined to the mental institution. He was trying to explain the sequence of events, but Nadia could hardly follow what he was saying. She kept on looking around expecting Melanie to show up, from behind the golden screen at the entrance to the cafeteria.

It was surely a practical joke. The young girls at the next table were counting their money carefully in order to pay their bill.

"I don't believe it." she said, but Hassan was not smiling.

"Nadia, your friend is seriously ill. She hasn't been well for some time now I'm afraid. I am very worried about her, especially because I will soon be leaving on a diplomatic mission to Syria."

"You mean you'll travel and leave Melanie in that place of horror?"

"I can't do anything for her now. She's in the hands of God."

"God? How long will the doctors keep her?" Nadia's voice caught in her throat.

"I don't know." His green eyes flashed with anger. "Did you know that your friend was a schizophrenic?" His voice was a notch too loud.

"What are you saying? No, I didn't know."

"She was seeing a psychiatrist and taking medications in secret. She never told you."

"No, I had no idea. How stupid of me, I didn't know she was suffering so much."

"Don't blame yourself; she's a good actress; she fooled us all."

"She loves you. Can't you see that Melanie kept her sickness a secret in order not to lose you?"

"She's destroyed my life."

"Your life? What about her life?" Nadia felt like hitting him. He was always so damned righteous.

"She was doomed from the start."

"I thought you were a religious man and believed in compassion for the weak."

"At this moment I wish she were dead. It would be easier on all of us."

"You disgust me." Nadia got up from her low chair. She felt dizzy and the turquoise walls surrounding her were making her sick. She left in a hurry, bumping her shin against the low coffee table, while trying with difficulty to negotiate her way out of the hotel. Her tears were blinding and her anger was disorienting.

Outside the hotel she could hardly breathe. The uniformed porter at the door of the hotel hailed her a rickety taxi and asked her where she was going. He took his time to slowly write down the destination and the license number of the taxi and finally gave her a copy of the slip.

As she tried to settle down in the dusty cab, she recalled the way Melanie used to be, with her wide smile and her love of life. She was the last person she thought of as mentally ill. True, she was crazy in her own way. Melanie used to gallop on her horse in the sand dunes behind the pyramids when all the other girls in class were afraid to canter. Was that crazy? She had dared to love without inhibition when her schoolmates had stayed virgins until their wedding day. Was that crazy? She was a brilliant dancer, who promoted dance as an art, in a society that considered it akin to prostitution. She broke the rules at every turn, did that make her crazy?

Nadia had confidence that her friend would get better somehow and face her life with strength. She opened her handbag and took out the acceptance papers to the University of California that Melanie had helped her get. She hadn't told her parents yet that she was planning to leave Sammy and Egypt. She wanted to face them at the last minute with all her plans in place in order to avoid arguments and recriminations.

She thought Melanie would be there to help her face the storm, but Melanie was in trouble. Melanie was in the Yellow Castle!

As the taxi was crossing *Kasr el Nil* Bridge, Nadia glimpsed a riverboat floating up the Nile. It reminded her of the time she had met Sammy on one of those boats, at the Sporting Club picnic.

"I'm not allowed to talk to boys," she had said.

"I'm not a boy, little girl." She was eighteen and he was twenty-four.

Then her thoughts returned to Melanie. She pictured her alone in the lunatic asylum among strangers and the thought made her break out into uncontrollable sobs. The taxi driver looked in his rear view mirror.

His thin dark face was startled.

"The world is full of grief, but you are too young and pretty to cry," he said.

She continued to cry for her friend and for herself. The taxi driver kept talking, trying to console her with a string of proverbs. "Every problem has a solution," he said, looking anxious as he waited to see the effect of his wisdom. "Every difficulty dissolves with time." He kept on trying, "Trust me, there will always be sunshine after the rain."

Nadia found herself finally smiling. The man flashed back a smile of relief and continued driving. She mustn't let life crush her. She was going to struggle and learn how to walk again, like she saw her father, who had suffered from a stroke and was half paralyzed, struggling every day to walk again. She must be brave for the sake of Ahab and the baby.

She must also be brave for Melanie.

The Yellow Castle seemed unchanged. It was much smaller than she had remembered, and the yellow color had faded from the sun and the dust to yellowish ochre. Nadia felt fear grip her heart as she entered the empty courtyard. There were no inmates roaming outside this time and the place seemed peaceful. She entered the lobby and asked about Mrs. Hassan Hosni.

"Just a minute, please," said a tired looking nurse.

While Nadia waited for twenty minutes on a lumpy armchair, she flipped through a magazine and marveled at all the new unrecognizable faces smiling at her from the glossy pages. The gossip columns about high society had vanished from the press, as well as the love stories by Ihsan Abd el Kuddus that all the teenagers had loved to read. The news was all censored and dictated by government policy. There were no more political cartoons making fun of the people in power. All opposition had been silenced and all literature served the ideals of the revolution. She found it all utterly depressing.

A solemn young doctor came up to Nadia and she got up to meet him. He stood hesitating for a minute before talking to her. "Excuse me, are you related to, to Mrs. Hassan Hosni?"

"I'm her friend."

"May your life be long; I'm sorry but your friend has passed away early this morning."

"What?" Nadia was shocked and sat down. She tried to get up but couldn't. She felt the ground giving way under her feet. Grief was tearing her guts and she just sat there crying. A nurse came forward and put her hand on her shoulder but the doctor waved her back. They stood aside and left Nadia to compose herself.

"I'm so sorry," said the Doctor. "It seems sudden, but we all knew that it was inevitable."

"Did you tell her husband?" The bastard must have already known.

"Yes, we informed him, he's on his way."

"*May God have mercy on her soul.* It's sad when someone so young goes before his time, but what can we do? *We all come from God and to Him we shall all return.*"

Nadia was not listening to what the Doctor was saying. The only thing she heard was that Hassan would be arriving at any moment and she definitely did not want to see him. She stood up to leave. She walked slowly towards the entrance.

"Melanie is dead," she kept repeating to herself trying to understand the meaning of the words. It was not fair. Melanie was cheated out of her life. She wasn't given a chance. She deserved better. She wanted to scream. "Melanie," she murmured. "I need you my dear friend. My dear Melanie, I am all alone without you, all alone" she cried. Suddenly she stopped. "What am I saying? No! I am not alone! I have Ahab to live for and the baby." Nadia took a deep breath and walked steadily towards the main door and out of the dark hospital into the bright sunshine.

28

ALEXANDRIA

THE SUMMER HEAT IN Cairo was unbearable. Nadia was impatient with all the delays and the bureaucracy involved in getting a visa to the United States. She agreed to go with her parents to Alexandria mainly because Ahab wanted so much to go. He would be delighted with the beach and with all the new friends he would encounter there. It would be a good holiday for them before the struggle she knew awaited her on all fronts.

Alexandria was linked to the best of her childhood memories. It was where she had spent most of her summers and school holidays.

Their building was right on the Corniche of Stanley Bay facing the Mediterranean Sea. The apartment was cool and the smell of the sea permeated the place.

The whole family was crowded into the apartment and she could hardly stand the lack of privacy. Nadia slept with her son in one bedroom, her sister-in-law slept with her three children in another, and her parents slept in the third bedroom. Her brother was working in Cairo and when he came on the weekends, he slept on a couch in the living room.

The family spent the whole day at the King's Muntaza Palace Beach where they had a cabin. The sandy beach surrounded a small bay and was covered by a sparse copse of pine trees. The Mediterranean here was gentle and transparent. She watched Ahab as he searched for sea urchins in the shallow waters of the bay. Ahab looked at Nadia, waved to her laughing.

She waved to him smiling, he was growing fast. He looked exactly like the pictures of Sammy when he was a little boy. As she watched him, all of a sudden her heart missed a couple of beats. She knew that instant that she loved him more than anything in the world.

Of course she had always loved him as any mother loves her child, but just now looking at him running around, laughing and playing, his skin turning brown in the sun, she realized that she would never love anyone else so completely.

She was glad that the neighbors in the next-door cabins watched her with alarm when she came out in her new flesh colored bikini. Nadia was only 26, had lost weight and looked like a teenager. The older women were shocked and the teenage girls were jealous. The men were eating her up with their eyes, but acted as if they didn't see her at all.

Let them stare. She didn't care for their good opinion any more. They were the same families who had shared that stretch of beach since it was opened to the public after King Farouk was exiled. Young people had married, new children were born, parents had aged and grandparents had passed away, yet everything was somehow the same, unchanged through the years.

She, on the other hand, had changed and they all felt it. It was not just the bikini. She was no longer one of them. The comfortable feeling of belonging wasn't going to be her destiny. She really didn't care as she was planning to leave the country forever.

They gave her sidelong glances and must have gossiped about Sammy whose escapades were always in the papers. She was aware that they even disapproved of her sundress with its thin straps and of the fact that she was traveling alone without her husband. To hell with them all.

She didn't seek their approval anymore!

She didn't join her mother and her friends when they played Rummy for hours on end. They gambled with pennies and felt sinful and wicked. She played baggammon with her father and infuriated him by winning almost every game. He never mentioned Sammy. He never forgave him for leaving Cairo for Beirut and leaving medicine for the movies. Her secret plans to leave the country helped her endure his anger and resentment. As for her husband, her love for him was slowly being erased from the walls of her memory.

Leafing through a magazine she saw the officers of the revolution standing in a group photo. They looked ten years older and twenty pounds heavier. They had all become ministers or ambassadors. She was startled to see Hassan standing beside the president at a meeting of the Arab League. Nadia hated him and would never forgive him. He had called her, but she had refused to see him.

"Don't blame me, Nadia. Melanie was ill. I didn't kill her," he said on the phone.

"Then why are you feeling guilty?"

"I loved her more than I loved life."

"May God preserve us from a love like yours. Your love was a curse," said Nadia. "I know you killed her as surely as if you had cut her throat, and I never want to hear from you again." She hung up the phone, feeling glad that she had a stab at him.

"Let's swear an oath that we will marry only for love," Melanie had once declared, and Nadia had not hesitated to swear too. It was a foolish oath to take in their world of prearranged marriages, but they were young and fearless. Melanie, who was so bright, had put all her faith in love, and it had destroyed her fragile existence. Yet, Hassan was not destroyed; he was still running around keeping busy with his career and being photographed with the president. As for herself, she who had never dreamt of a life without the only man she had ever loved, was about to try and learn to live on her own.

In spite of her sadness, the laughter of the children, the taste of good food, and the extended summer days lulled Nadia into a content mood while she waited patiently for her U.S. visa.

One evening, while the family was gathered in the living room, in front of a new television set, watching an old *Abd El Wahab* movie, she ran to answer the phone that was ringing. Obeid's voice came indistinctly over the crackling line.

"Sammy's in jail, you have to come quickly." She felt the ground sway under her feet.

"What? What happened?" She supported herself on the delicate table on which the phone was sitting.

"He and Romy were arrested for drug possession."

"Who? What? With Romy? I don't understand?" she said softly. She turned her back to the room, trying to avoid being overheard by the rest of the family. Her voice was shaking.

"They claim he's a drug dealer," Obeid was saying.

"Oh God . . . that's ridiculous."

"I know. Just come over as fast as you can. You've got to get a lawyer to get him out."

Her heart was beating so loud she could hardly hear what Obeid was saying. Of course, she had to help Sammy. Did she still love him?

In spite of everything? She was utterly confused, but she had to go.

"Yes . . . yes Obeid, I'll be there as soon as possible." Her voice was breaking. The rickety table she was leaning against squeaked and threatened to collapse as she hung up. 'Oh Sammy Sammy, why are you so irresponsible?' she thought.

She remembered the evening in Deir el Qamar when Sammy was smoking hashish with the actors and the crew after a day's work. She had entered a room filled with smoke. They were all reclining on pillows; some were smoking the *narguila*, others were smoking cigarettes. She had sat down beside Sammy on one of the pillows, and Romy had passed her a joint.

"I can't smoke. It makes me sneeze," she had said.

"Don't be silly, take a puff," Romy had urged.

"It's wasted on me, I can't inhale but I still like the smell."

They had laughed at her and dismissed her as annoyingly conventional. She had felt that she was immediately excluded from the circle. When she had, later on, expressed her concern, Sammy only laughed and told her that it was perfectly safe even if it was technically illegal. It was a normal thing to do; in Lebanon, everybody did it.

Before she left for Beirut, she needed a letter from her husband consenting to let her travel. She had to ask her brother to write her the letter. He saw her determination and did not question the reason for her request. He forged Sammy's signature without hesitation. Even though she hadn't told him what was happening, he had guessed that there was a problem and that she had to leave urgently.

Her mother was happy to have Ahab stay with her for a few more weeks. Nadia lied to everyone; she invented a story about Sammy being taken ill. They accepted her story with ease. She felt that nobody was really interested in the reason for her sudden departure. It was easy for them to let her go because she was no longer their responsibility. They believed that she belonged to Sammy and now depended on him. If only they knew.

29

FACING THE MUSIC

ONCE NADIA ARRIVED IN Beirut, she found Obeid waiting for her at the airport. He was driving one of Mansour's cars.

"Sammy said to talk to Mansour. I think he would be willing to help."

Of course Nadia knew why Mansour would be willing to help, but she didn't want to get involved with him. She called the singer Nagwa, who had been a good friend of Sammy's, and asked her for the name of the best lawyer in town. Nagwa was forever in trouble and had many brushes with the law. She was frequently being swindled by her various lovers and seemed to always forget to pay her taxes. She would certainly know the name of a good lawyer.

"Oh my God! I'm so sorry for what happened," she told Nadia. "I would do anything to help Sammy; he is such a nice man, I swear to God. I owe him so much. I swear that the film he directed for me was the best movie I've ever made." Nagwa then suggested El Assy. "I swear to God he's the best lawyer in town. I'll call him right away."

Nadia had no choice but to put her faith in Nagwa. She had no one else in Beirut that she could depend on.

The lawyer, Maher El Assy, had very luxurious offices in the center of Beirut, for his clients were only the rich and famous. He was in his early forties but had already gained the reputation of being honest and efficient.

He slowly weighed every word when he spoke and asked her probing questions. He had a calm reassuring demeanor, and he quickly put her at ease. He assured her that he would get Sammy released in a very short time.

El Assy was of medium height, slim and elegant. Nadia hadn't noticed that he had only one arm until she first tried to shake hands with him and he put out his left hand. She was confused and blushed when she saw that he had an empty sleeve hanging lifelessly from his right shoulder. Nadia liked him, even after he asked her for an exorbitant sum of money for his fees. She told him she would have the money in a few days.

She really didn't want to ask Mansour for the money. She knew he would be difficult to handle. He had been pursuing her from the first time he saw her, but the problem now was that she was also very attracted to him. It was asking for trouble. That situation was a minefield threatening to explode. Nadia didn't need more complications at this moment in her life.

There was no question of asking her parents for help, for she had decided to keep them in the dark, even though they might read about the disaster in the papers. Now she needed to raise some money quickly.

Nadia still had her jewelry. It was sitting in the safe in her closet, rarely used. She had planned to sell it to finance her trip to America, but America now seemed to be so far away, just a dream. Resolve one problem at a time, she told herself. She decided to sell something right away to pay the lawyer. Her mother had a Lebanese friend whose husband was a famous jeweler. She looked her up in the telephone book and forced herself to call her. She fought the feelings of shame, as she explained to Nazly Hanem the situation, and asked the woman if she could help her sell a diamond bracelet.

"*To'burini*, may you bury me," Nazly Hanem had said. "Come over to my house at once. I'll see what I can do."

Nazli Hanem was a corpulent lady, whose milk-white arms were covered with golden bracelets and who smelled of rose water and cinnamon.

"Please, I don't want my parents to know what happened, I want to resolve this problem myself."

"Of course, *Ya habibti*, your secret's in a deep well my love," Nazly Hanem reassured her.

Nadia couldn't bear to think of what her mother would say if she found out. She could hear her voice shouting in her ear. "We warned you. He's irresponsible. We told you so." She would bombard her with recriminations and accusations. "How can I show my face? You have disgraced us." Nadia had heard all those laments many times before and didn't need to hear them again. She would be reminded over and over again that he was her choice and that she had brought shame on the family. No, her parents should never find out about Sammy's mishap, if she could help it.

Nazly Hanem promised to call her the minute she found a client.

Nadia stayed alone in the Beirut apartment. It was so silent and empty without Ahab and Sammy. She waited for a phone call from El Assy to give her news of the situation. It was eight o'clock in the evening when the doorbell rang. She put on her robe and went to investigate.

She saw through the peep hole three men standing patiently at the door. She was sure they must be looking for someone in the next apartment. "Who is it?" she asked.

"Police. Please open the door."

She hesitated for a second, and then opened the door a sliver. The largest of the three men put his foot in the door and showed her his badge.

"May we come in?" he said.

"I'm all alone; can you please come back in the morning?"

He pushed the door open and stepped in. "No, we have to come in now," he laughed in an unpleasant way. The other two followed him, looking around as if they expected to find someone else in the apartment.

"What do you want?"

"We would like to search the apartment," said the giant officer, smiling.

"Do you have a search warrant?" she asked. She forgot that she was in Beirut and not watching an American detective film.

They stared at her in disbelief and disregarded her question. They were like characters in a movie. She labeled them, the big man, the bald man and the skinny man. They were all dressed in nondescript gray suits and white shirts. The skinny one had a rumpled tie around his neck and was always smirking.

"Stay where we can see you," said the bald one, giving her his back.

She could see a gun in his back pocket. She was starting to shake. Her stomach twisted and she felt a sudden urge to go to the bathroom. She had her jewelry in the safe and was afraid the policemen would take it away from her. "Can I call my lawyer?"

"Sure, go ahead," replied the skinny man who was standing close to her. He had terrible bad breath. She felt like slapping him to wipe that permanent smirk off his face. Luckily, he turned away and started opening the closet doors in the entrance.

She went into the bedroom, to find the number of the new lawyer, and remembered that Sammy had put his stash of hashish in the top drawer of her dresser. She started to sweat with fear. She managed to close the door of the bedroom softly, opened the drawer quickly, and in a panic riffled through her clothes for the small package wrapped in cellophane. There was nothing there. Relieved, she sat on the bed to collect her thoughts and compose herself. The big policeman opened the door and rushed in.

"What are you doing?" he asked in a loud and angry voice. His suit was rumpled and she could smell his sweat.

"Just looking for the lawyer's number. I can't remember where I put it." She fumbled in the drawer of the side table beside the bed.

"Leave that door open and hurry up." He stood there staring at her, refusing to move.

She pretended to find her telephone book, which had always been right there in front of her. By now she was shaking visibly. After several unsuccessful attempts, she dialed the number correctly. The phone rang several times before El Assy finally picked up. She told him that the police were in the apartment.

"Do they have the right to barge in like this?" she asked, looking at the big man who was staring at her. "He wants to talk to you," she told the big man and gave him the phone. They talked for a few minutes in whispers and then the policeman held the phone to her. His expression had softened a little.

"Don't worry. Let them search. It's just routine," El Assy said.

Nadia was pleased because she felt that El Assy was sure she had nothing to hide.

The three men searched every inch of the apartment. She had placed empty crates around the place to use as tables until they could afford real furniture, and had covered them with pretty pieces of material and tablecloths. These boxes intrigued the policemen, and they turned them over and over in their fruitless search. Finally they left.

She couldn't sleep after that and kept walking around the apartment all night, searching in the places that they had overlooked for she knew not what. She was being hysterical. The sight of Sammy's clothes neatly hanging in the closet seemed incongruous with the turmoil that had turned their lives into an untidy mess.

Dealing with her personal disaster was now a priority. She told herself that she had to be strong and clearheaded. Do not panic, she begged the new person she had suddenly become.

The next morning, she was expecting the lawyer to call but was surprised when Mansour rang her doorbell. She was just coming out of the shower, dressed in her bathrobe. She opened the door and he entered.

"When you didn't call, I said I'll pass by you. I'm sorry to disturb you, but I thought you might need help," he said.

"Thank you," said Nadia adjusting her robe. "Nagwa introduced me to a good lawyer."

"You will need some money," Mansour said, devouring her with his eyes.

"I think I will be okay."

"Nadia, I hope you will consider me as a friend." He came close to her and put his hand on her shoulder. She almost took a step back, but hesitated. Mansour took her hesitation as a signal of acquiescence. He enfolded her in his arms and started kissing her neck and working his way towards her face and mouth. They kissed ardently. He was a good kisser. Soon one of his hands had found her breast and started fondling it. She was sexually aroused. When he opened her bathrobe and started slowly to slip his hand between her thighs, she regretted her weakness and pushed him away.

"I'm sorry," he said. "Please forgive me. I'm crazy about you."

"Mansour, please go away. I have enough problems at the moment."

"I'll always be here if you need anything. Don't worry. I will not demand a pound of flesh in exchange for my services. I'm not as bad as you think."

Nadia laughed nervously. "I know. I appreciate your offer."

"Nadia, may I ask you a question? Why did you kiss me?"

"Because I'm crazy about you too," she said.

He laughed. "Then, why are you sending me away?"

"Because I don't need another womanizer in my life."

Mansour smiled. "Fair enough," he said, and started for the door.

At the door, he stopped and turned. "By the way, you need not worry about money. I have already paid El Assy."

Nadia smiled. "I knew you would. Now Sammy can finish your film."

Mansour laughed out loud. He knew that he had met his match.

CHILDREN OF THE NILE

T HE COURTHOUSE WAS SITUATED in the center of old Beirut, near Liberty Square. It must have been built in colonial times. Nadia had never liked Beirut; it didn't compare to the grandeur of Cairo, even if it was becoming the so-called Switzerland of the Middle East. It was a city of small shopkeepers and greedy bankers.

She arrived at the courthouse sweaty and exhausted. The humidity and heat of Beirut in August made her irritable, and the traffic had been more unbearable than usual. She entered the old stone building. The men standing around in the corridors of the courthouse eyed her with frank appraisal. No one bothered with discreet glances in this town.

She observed everything around her in a detached manner. She was a spectator in an unfamiliar drama, and she watched and waited for the action to unfold. She met El Assy, and together they sat in the huge courthouse hall, waiting for permission to see the prisoner.

"This is a very interesting building," she said, looking around her at the marble floor and classic columns.

"It's the French influence of the nineteenth century. As you must know, Lebanon was under French control since 1860 when they intervened to save the Catholic Maronites from being massacred by the Druses. The result of the French occupation is that we are still under Napoleonic law, which means a person is guilty until proven innocent. It means that Sammy has to stay in jail until we can prove his innocence."

Nadia's heart missed a beat. "I understand," she said.

El Assy excused himself for a minute to talk to the district attorney.

She was left alone with her thoughts for a long while.

She had loved Sammy from the time she was a teenager, but that was a long time ago. Did she still feel the same about him? She was not sure.

She had always stood by him through thick and thin. Life with him was like a crazy ride on a rollercoaster, up and down, up and down. But after a while, one gets tired. She never complained through his change of careers, through his love affairs, and most probably he thought that her acceptance was permission for his philandering. Should she have complained? Should she have yelled? Should she have given him an ultimatum? Well, it's too late now and here they were.

The Lebanese papers had published pictures of Sammy and Romy on the front page. The headlines boldly stated, "International Drug Trafficker Apprehended." She had never felt safe in Lebanon; it was a lawless Mid-Eastern version of the Wild West. All men were armed and they shot in the air in jubilation, and sometimes at each other, at the slightest provocation. On the other hand, Egypt had become the last place in which she wanted her children to grow up. The revolutionary government was floundering, trying to impose, without great success, a brand of socialism imported from General Tito in Yugoslavia.

She had hoped to immigrate to the United States. America was a young and rich country where one could wake up in the morning with a fair chance of still having a responsible government. Wars always took place in faraway lands, like Korea or Vietnam, and the dollar was stable.

If she had a decent job there, she could provide for her children. But what was she going to do now?

Finally, after what seemed like hours of waiting, El Assy and a police officer escorted Nadia to the visiting room at the back of the building.

In the small room, where they took her, she found Sammy sitting alone on a bench on one side of a long wooden table. An armed guard was standing behind him. He had aged ten years in those two weeks. The light coming in through two small windows reflected the shadow of the iron bars on the bare cement floors. The guard motioned to her to sit across from Sammy and El Assy left the room to give them some privacy.

"I'm sorry, I'm sorry, I've ruined everything. Please forgive me. I'm sorry," he started to cry. Nadia noticed that many white hairs had sprouted on his head. She felt sorry for him and for herself, but her anger kept her from breaking down. It gave her courage.

"Sammy, why the Hashish and why Romy? She's married, plus she's your boss' mistress for heaven's sake."

He stopped crying, "Maybe that's why I was attracted to her."

"What? You wanted to show him that you're more of a man than he was?"

"I don't know, maybe. Please forgive me Nadia, I'll change, I swear I'll change."

She had heard that before. He will never change.

"I'll be a different man. Trust me." He started to cry out loud.

"Be a man," she told him angrily, "It's too late for tears. We have to talk about what we must do to get you out of here."

"Yes, you're right," he said. He stopped crying instantly. "Can you forgive me?" His eyes met hers for a moment. She could have weakened, but she didn't.

"The lawyer says that if you had only a small amount of the stuff for personal use, you might get a very light sentence."

"I didn't have any hashish on me at all. Romy had some in her handbag," Sammy started whimpering again. He seemed to be a stranger to her. This was the man who, only weeks ago, was her reason for living.

"Don't worry, you'll be all right."

"How long do you think it's going to be?" he asked.

"You'll be out very soon." She didn't tell him it could take years if the judge was inclined to teach him a lesson.

"I don't deserve you; you've always been my greatest ally and support," he continued. He held his hands together with his fingers intertwined as if in prayer. He seemed to be shaking all over.

"Sammy, what you did isn't considered a serious crime. It'll take some time, but you'll be out."

"Thanks for coming to help me," he murmured.

"What did you expect? You're the father of my children, after all."

"How is Ahab?"

"He's fine. I left him with my mother."

"What did your parents say?"

"They don't know, I didn't tell them."

"I knew that I could count on you. I promise that I'll be the best of husbands and never give you any more cause for pain. You'll see, I promise," he told her.

She looked down at the table that separated them. "I promise," he had said. Nothing changed. How many times did he promise? How many times did he say he'll change? She couldn't envision a future for them. With him she would always be the faithful wife, supportive and forgiving. She would always be half a person living half a life.

"They will not break me. You'll see. I'll come out and be the greatest director in the world."

That was the old Sammy she knew. She had loved his courage and sense of adventure. But no more. No more. She wanted to live her own adventure now.

"Is he a good lawyer?" Sammy asked.

"He's the best," she said.

"Ask Mansour to give you some money. I'll pay him back."

"Don't worry, he already paid the lawyer."

"Good. You know, I'm sure it was Mansour who tipped the police. He was very upset that Romy preferred me to him."

"Who cares who denounced you?" Nadia said angrily.

"Hey, don't get angry. You handled a lot of my shit before."

"Yes, but no more my dear Sammy, no more. In fact, I've decided to leave you."

"What? Are you crazy?" He looked surprised. "You can't leave me now."

"You're the one who left me, remember?" she said.

"It's all over with Romy; it was just a crazy infatuation."

"Sammy, I don't give a shit about Romy! If it isn't Romy, it would be another pretty woman."

"That's not true . . ." Nadia interrupted him.

"I need to make sense out of my life, and my children need stability. Actually, what I really need is just simply to have a life."

"So you're deserting me now, when I'm in jail, helpless?"

Sammy was acting again, she thought.

"I told you I'll take care of things and get you out of here, but I want you to promise to let me go." Nadia knew that Sammy had the legal power to detain her from leaving the country. It would be impossible to get a divorce or have custody of the children without his consent.

"I'll never let you go. You're mine and will be mine forever." He banged on the table.

"Calm down," said the soldier.

"Sammy, I swear I'll let you rot in jail for life if you don't promise to let me go." She said forcefully. It was the wrong time for him to be threatening her.

"You know that you'd never do that to me," he said disturbed.

"Really? Think about it in your cell."

"Wow, what happened to you in a few weeks?"

"It's not just a few weeks. It took way too long for me to wake up."

"What about our love?"

"Our love? What about it? I was weaned on tales of love. I was told the wrong stories and I believed in them."

"Nadia we can talk about this later," Sammy said, with the ghost of his crooked smile appearing on his face. Nadia felt a twinge of surprise, for she suddenly knew that she was no longer under his power. She had broken loose and no sweet talk was going to change her mind. The old Nadia was gone forever. She fixed her gaze on the barred window behind him.

"Yes," she answered him calmly as if she were talking to a child; "We can talk about this later but I will not change my mind." She was no longer afraid of him.

Their time was up. She stood up and whispered goodbye. She had to turn away quickly because tears were filling her eyes.

Outside the courthouse, she waited to say goodbye to El Assy. He came out laughing aloud.

"What happened?" Nadia asked him.

"Imagine the district attorney told me he found out that Sammy was trying to convince two of the biggest drug dealers in jail to invest in his next film."

Nadia laughed, "They will never break him. Sammy will always be the same. He is a survivor," she told El Assy.

"I think you're right. I will see you in court," he said as he turned away and left.

Sammy will never change, but she had definitely changed. The weather had mercifully turned cooler. She decided to walk all the way across Beirut to her apartment. Nadia took a deep breath. She felt like a new person. She was finally free, and America was not a dream anymore.

www.ingramcontent.com/pod-product-compliance
Lightning Source LLC
Chambersburg PA
CBHW031501120626
46545CB00005B/1691